Magnetic Resonance Elastography

Sudhakar K. Venkatesh

Richard L. Ehman

Editors

Magnetic Resonance Elastography

Springer

Editors
Sudhakar K. Venkatesh, MD
Department of Radiology
Mayo Clinic College of Medicine
Mayo Clinic
Rochester, MN, USA

Richard L. Ehman, MD
Department of Radiology
Mayo Clinic College of Medicine
Mayo Clinic
Rochester, MN, USA

ISBN 978-1-4939-5095-9   ISBN 978-1-4939-1575-0 (eBook)
DOI 10.1007/978-1-4939-1575-0
Springer New York Heidelberg Dordrecht London

Printed on acid-free paper

Springer is part of Springer Science+Business Media (www.springer.com)

Dedicated to my parents, K. Venkatesh Rao and Chandrakala
And my Mentors and Teachers for their inspiration, guidance
and support

Sudhakar K. Venkatesh

Foreword

Physicians have used their sense of touch as a powerful diagnostic tool for centuries. It has long been known that many disease processes are associated with major changes in the mechanical properties of tissue. Historically, investigators in the fields of biomechanics and physiology have sought to integrate our knowledge of the mechanical properties of normal and abnormal tissue into our understanding of biological function. Building on these advances, we have more recently seen the emergence of the new science of mechanobiology. This has revealed the profound extent to which forces and physical properties of the cellular environment can affect cellular behavior. Indeed, it is now known that abnormalities in the mechanical environment contribute to the development of many diseases through a process known as mechanotransduction.

Motivated by these considerations, researchers have worked to develop noninvasive technologies for quantitatively evaluating the mechanical properties of tissue. A very promising approach that has emerged from this effort is Magnetic Resonance Elastography. This innovation, developed with the support of the National Institute of Biomedical Imaging and Bioengineering at the National Institutes of Health, makes use of acoustic and magnetic resonance phenomena to accurately and noninvasively identify the disease. Over the last few years, Magnetic Resonance Elastography has been successfully translated from the laboratory to clinical practice. Even at this early stage, it is now used in medical care worldwide.

I am delighted to write a foreword for this, the very first book on MR Elastography. The book starts with a history of Elastography, providing the readers an overview of the development of the technique. The following chapters provide coverage of currently established applications and a preview of things to come. This historic book would be useful for radiologists, imaging scientists, technologists, residents, and fellows who are interested in learning about the new and exciting field of MR Elastography.

Roderic I. Pettigrew, PhD, MD
Director
National Institute of Biomedical Imaging
and Bioengineering (NIBIB)
Bethesda, MD, USA

Contents

Contributors

Philip Araoz, MD Department of Radiology, Mayo Clinic College of Medicine, Rochester, MN, USA

Sumeet K. Asrani, MD, MSc Baylor University Medical Center, Dallas, TX, USA

Sabine F. Bensamoun, PhD Laboratoire de Biomécanique et Bioingénierie, Université de Technologie de Compiègne (UTC), Compiègne, France

Jun Chen, PhD Department of Radiology, Mayo Clinic College of Medicine, Mayo Clinic, Rochester, MN, USA

Richard L. Ehman, MD Department of Radiology, Mayo Clinic College of Medicine, Mayo Clinic, Rochester, MN, USA

Kevin J. Glaser, PhD Department of Radiology, Mayo Clinic College of Medicine, Mayo Clinic, Rochester, MN, USA

Roger C. Grimm Department of Radiology, Mayo Clinic College of Medicine, Mayo Clinic, Rochester, MN, USA

Rolf D. Hubmayr, MD Mayo Clinic College of Medicine, Mayo Clinic, Rochester, MN, USA

John Huston, III, MD Department of Radiology, Mayo Clinic, Rochester, MN, USA

Arunark Kolipaka, PhD Department of Radiology, Columbus, OH, USA

Department of Radiology, The Ohio State University, Colombus, OH, USA

Jennifer Kugel Department of Radiology, Mayo Clinic College of Medicine, Mayo Clinic, Rochester, MN, USA

Diagnostic Radiology Research, Mayo Clinic, Rochester, MN, USA

Armando Manduca, PhD Department of Radiology, Mayo Clinic College of Medicine, Mayo Clinic, Rochester, MN, USA

Yogesh Mariappan, PhD Philips Healthcare, Bengaluru, Karnataka, India

Kiaran P. McGee, PhD United States Naval Research Laboratory, Washington, DC, USA

Anthony Romano, PhD United States Naval Research Laboratory, Washington, DC, USA

Jayant A. Talwalkar, MD, MPH Division of Gastroenterology and Hepatology, Mayo Clinic, Rochester, MN, USA

Sudhakar K. Venkatesh, MD, FRCR Department of Radiology, Mayo Clinic College of Medicine, Mayo Clinic, Rochester, MN, USA

Meng Yin, PhD Department of Radiology, Mayo Clinic College of Medicine, Mayo Clinic, Rochester, MN, USA

Introduction

Sudhakar K. Venkatesh and Richard L. Ehman

Magnetic resonance elastography (MRE) is a powerful technology for quantitatively imaging the mechanical properties of tissue that was invented at the Mayo Clinic and first described in a publication in SCIENCE [1]. A special MRI-based technique was developed, capable of imaging propagating mechanical waves within the living human body, even though they displace tissue by just fractions of microns. Advanced mathematical techniques are used to process these data to generate quantitative images depicting the stiffness of tissue.

MRE has shown a capability to noninvasively quantify the mechanical properties of many tissues including liver, muscle, brain, lung, spleen, kidneys, pancreas, uterus, and thyroid. The first well-established clinical application of MRE is the assessment of liver fibrosis. MRE has been routinely used for this purpose in patient care at the Mayo Clinic since 2007. For many patients, MRE is a reliable, safer, more comfortable, and less expensive alternative to liver biopsy for diagnosing liver fibrosis. MRE has been commercially available as an FDA-cleared upgrade for MRI systems since 2010. At the time of writing, MRE technology is available from the manufacturers of more than 90 % of the MRI scanners in use around the world.

Now that MRE technology is in the hands of the clinical and scientific community worldwide, we anticipate that many new areas will be explored and innovative clinical applications established. Ongoing technical research will focus on many opportunities to address challenges relating to pulse sequence design, delivering mechanical waves to all areas of the body, and improving processing algorithms to generate more accurate and higher images. MRE is an exciting field with scope for innovations, technical improvements, and new clinical applications.

In this book, we present a perspective on the rationale and origins of elastography, and then a series of chapters describing the emerging applications of MRE in various organs and clinical settings. We hope that this book addresses the gaps that exist in understanding the principles and applications of MRE as well as emerging clinical applications. The book is a result of hard work by all the authors and we sincerely hope that readers will find it to be a useful introduction to the field.

Reference

1. Muthupillai R, Lomas DJ, Rossman PJ, Greenleaf JF, Manduca A, Ehman RL. Magnetic resonance elastography by direct visualization of propagating acoustic strain waves. Science. 1995;269(5232):1854–7.

S.K. Venkatesh, M.D., F.R.C.R. (✉) • R.L. Ehman, M.D.
Department of Radiology, Mayo Clinic College
of Medicine, Mayo Clinic, 200 First Street SW,
Rochester, MN 55905, USA
e-mail: venkatesh.sudhakar@mayo.edu;
ehman.richard@mayo.edu

S.K. Venkatesh and R.L. Ehman (eds.), *Magnetic Resonance Elastography*,
DOI 10.1007/978-1-4939-1575-0_1, © Springer Science+Business Media New York 2014

Kevin J. Glaser and Richard L. Ehman

Introduction

Diagnostic medicine involves a synthesis of information about the current and past health of a patient which is used to diagnose the patient's condition and to design a plan for future medical care and treatment. This information can come from many sources, including discussions about family medical history to indicate hereditary conditions, the results of blood and culture tests to indicate the presence of known pathogens and to assess the chemical and cellular balance of the patient, and the results of imaging tests (e.g., X-ray and MRI) to determine abnormalities in the structure and function of particular organs. A common technique used in routine physical examinations is palpation [1, 2]. Among its numerous uses, palpation allows physicians to assess changes in the mechanical properties of tissue associated with the presence and development of certain diseases. Breast and thyroid cancer, for example, are often detected as a stiffening of normally soft glandular tissue [3]. Similarly, liver disease often results in cirrhosis

of the liver, a condition recognized by a marked increase in the stiffness of the entire liver [4].

The power and utility of palpation can be seen in reports that many breast cancers are detected by palpation before being found using mammography, and some cancers detected using palpation are even occult on mammography [5–7]. Similarly, abdominal tumors are often found during surgery that were not detected earlier by CT, MRI, or ultrasound [8]. While palpation continues to be an important tool for clinicians, it is still a qualitative technique that is limited to tissue accessible by the physician. The use of imaging techniques designed to quantitatively assess the mechanical properties of tissue, even tissue not directly accessible by touch, could provide significant information to physicians. Since there are known changes in the mechanical properties of tissue associated with the advanced stages of many diseases, it is likely that there are changes to tissue that occur at even earlier stages of these diseases that may be useful for diagnosing the disease, predicting the course of the disease, and monitoring the effect of treatment.

Modern medical imaging has become an invaluable tool for assessing the structure and function of healthy and diseased tissue in vivo and noninvasively. For example, techniques such as film radiography, CT, MRI, PET, SPECT, and ultrasonography can provide information about bone fractures, tissue degeneration, abnormal blood flow and perfusion, and tumor location and margins. Each of these techniques relies on

K.J. Glaser, Ph.D. (✉) • R.L. Ehman, M.D.
Department of Radiology, Mayo Clinic College
of Medicine, Mayo Clinic, Rochester, MN, USA
e-mail: Glaser.Kevin@mayo.edu;
Ehman.Richard@mayo.edu

S.K. Venkatesh and R.L. Ehman (eds.), *Magnetic Resonance Elastography*,
DOI 10.1007/978-1-4939-1575-0_2, © Springer Science+Business Media New York 2014

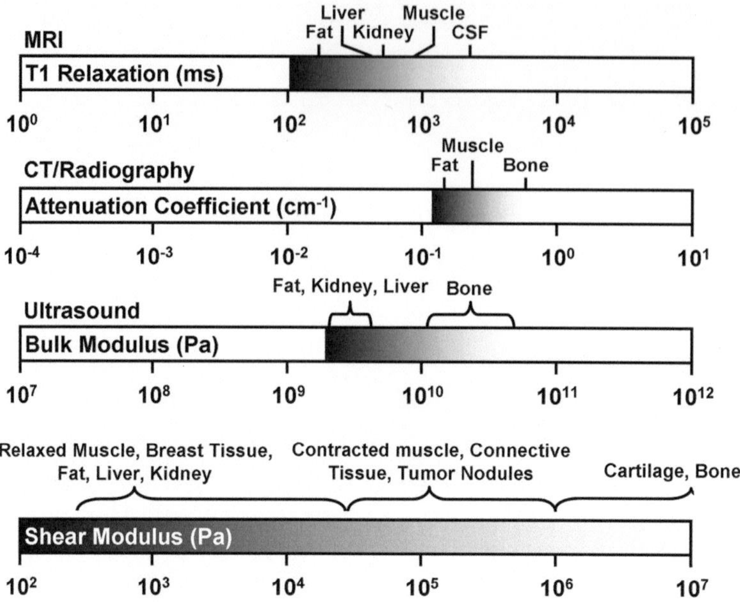

Fig. 2.1 Examples of tissue properties responsible for contrast in imaging techniques including MRI, CT, and ultrasonography. The T1 relaxation rate, X-ray attenuation coefficient, and bulk modulus of soft tissues vary over about one order of magnitude. The shear modulus, a measure of tissue stiffness, varies by several orders of magnitude for healthy and pathologic tissues

differences in specific properties of tissue to provide contrast in the images they produce. For example, while MRI uses differences in tissue magnetic properties, CT and film radiography produce images based on differences in the absorption of X-rays by different tissues. PET and SPECT rely on measuring the decay of radionuclides differentially absorbed by healthy and diseased tissue, whereas ultrasonography is based on detecting differences in the acoustic impedance (e.g., density or bulk modulus) between various tissues.

The properties of tissue involved in some typical radiographic imaging methods are shown in Fig. 2.1 and can be seen to vary by only about one order of magnitude, even between healthy and diseased tissues. This can limit the contrast and detectability of some tissues and structures. From the mechanical testing of ex vivo tissue samples, it has been shown that some tissue mechanical properties (e.g., the shear modulus) can vary over several orders of magnitude between different types of tissue and between

normal and diseased tissues [9, 10]. Therefore, an imaging technique capable of measuring tissue mechanical properties could be quite sensitive to disease-related changes in these properties.

Elasticity Theory

Measuring and modeling the mechanical properties and behavior of materials is an important part of many fields of science. From improving the design of roads and buildings to creating artificial tissues and organs, knowledge of the mechanical properties of the individual components of a system and of the system as a whole are critical for the adequate performance of the final structure. Considerable research has been involved in the development of artificial tissues and organs, such as the heart, lung, liver, skin, and cartilage [11–14]. In some cases, the primary goal is to replicate the function of the original organ (e.g., the artificial heart). However, in other cases, reproducing the mechanical properties of the original tissue is equally important because reproducing the mechanical response of the tissue under physiologic conditions will increase the chance that the patient will be able to return to normal activities with a normal range of motion and function. Therefore, understanding the mechanical properties of healthy and diseased tissue is important not only for diagnosing and characterizing a disease, but also for understanding and monitoring tissue function.

A number of mechanical tests exist for measuring the static and dynamic properties of materials (e.g., indentation, impact, creep, and torsion tests) [15–18]. These mechanical tests involve applying a known force (or stress) to the material and measuring the resulting displacement (or strain) of the material. For example, in an indentation test, a small probe is used to deform a small sample of a material. By measuring the amount of force that is being applied by the probe and the amount of deformation of the sample that results, material properties such as the Young's modulus, another indicator of material stiffness, can be determined. The response of a material during a particular test will depend on several mechanical properties, including anisotropic, nonlinear, thermal, geometric, composite, plastic, viscous, flow, and elastic effects. By carefully controlling the design of these experiments, and by making simplifying assumptions about the characteristics of the material and the applied stresses, the results of these measurements can be interpreted in meaningful ways (e.g., [17–31]).

A typical set of assumptions made when measuring the mechanical properties of a material is that of an infinite, homogeneous, linear, viscoelastic material [9, 32–34]. This set of assumptions significantly reduces the complexity of the mathematical relationship between the applied stresses and the resulting strains, and also reduces the number of unknown material properties that have to be determined to characterize the material. By assuming the material to be infinite, the impact of certain boundary conditions and boundary effects can be ignored. Assuming the material to be homogeneous (at least within a small region where data processing is performed) means only one set of material parameters needs to be considered and changes in the mechanical properties in space are negligible. Treating the material as linearly viscoelastic means that the strain of the material is linearly related to the applied stresses, though the response may change depending on the rate at which the stress is applied. Clearly, the validity of assumptions such as these must be reconsidered with every application because different materials and different experimental setups will have different characteristics.

Under the above assumptions, a fully anisotropic material may contain 21 independent quantities that would have to be known in order to fully characterize the material. The number of unknowns can be reduced by assuming that the material has certain symmetries. In a transversely isotropic material, the material is considered to have one preferential direction in which the response of the material is different from the response in the orthogonal directions. For example, in muscle, the response of muscle tissue along the direction of the muscle fibers to a particular stress will be different than the response of the tissue to a similar stress applied across the muscle fibers. Transversely isotropic materials can be described using as few as three parameters. Another common symmetry assumption is that the material under investigation is isotropic, and thus responds equally to stresses in any direction. An isotropic material only requires two quantities to describe its mechanical behavior, and several such quantities have been defined to aide in the description of the behavior of isotropic materials. These quantities include the Lamé parameters λ and μ (μ also being called the shear modulus), Poisson's ratio (ν), Young's modulus (E), the bulk modulus (K), and the P-wave modulus (M). Knowledge of any two of these quantities allows for the calculation of the others. For example, by knowing the shear modulus and Poisson's ratio, the others can be calculated as

$$E = 2\mu(1+\nu), \quad \lambda = \frac{2\mu\nu}{1-2\nu},$$

$$K = \frac{2\mu(1+\nu)}{3(1-2\nu)}, \quad M = \frac{2\mu(1-\nu)}{1-2\nu}. \tag{2.1}$$

To facilitate the analysis of the mechanical response of different types of materials under different loading conditions, the equations relating the stresses and strains can be expressed in different forms [9]. One such technique is to study the frequency-domain equations of motion for an infinite, isotropic, homogeneous, linear, viscoelastic material experiencing time-harmonic motion (such as a periodic deformation of tissue) [33]. These equations can be written as a set of complex-valued, coupled differential equations as

$$-\rho\omega^2 U(r,f) = (\lambda + \mu)\nabla(\nabla \cdot U(r,f))$$
$$+ \mu\nabla^2 U(r,f)$$

or

$$-\rho\omega^2 U(r,f) = (\lambda + 2\mu)\nabla(\nabla \cdot U(r,f)) \qquad (2.2)$$
$$- \mu\nabla \times (\nabla \times U(r,f)),$$

where ρ is the density of the material, $\omega = 2\pi f$, f is the frequency of the harmonic motion, and $U(r, f)$ is the vector displacement of the material at the position r. These equations can be used to show that, under the above assumptions, the response of the material is to propagate shear waves and longitudinal waves. Longitudinal waves have the property that the direction of the displacement of the material is in the same direction as the direction of wave propagation. Shear (or transverse) waves have the property that the displacement is perpendicular to the direction of wave propagation. The wave speed of the shear and longitudinal waves (c_S and c_L, respectively) can be written as

$$c_S = \sqrt{\frac{2|\mu|^2}{\rho(\Re(\mu) + |\mu|)}} \xrightarrow{\Im(\mu) \to 0} \sqrt{\frac{\mu}{\rho}},$$

$$c_L = \sqrt{\frac{2|\lambda + 2\mu|^2}{\rho(\Re(\lambda + 2\mu) + |\lambda + 2\mu|)}} \qquad (2.3)$$
$$\xrightarrow{\Im(\lambda + 2\mu) \to 0} \sqrt{\frac{\lambda + 2\mu}{\rho}},$$

where $|\mu|$, $R(\mu)$, and $I(\mu)$ indicate the magnitude, real part, and imaginary part of the complex-valued quantity μ, respectively.

It has been shown that the shear wave speed can vary significantly between different healthy and diseased tissues, from less than 1 m/s to over 100 m/s [10, 35, 36]. However, the longitudinal wave speed for most soft tissues is about 1,540 m/s and does not vary significantly between different types of tissue. Therefore, assessing the shear modulus and the shear wave speed has been the primary target for researchers trying to use the mechanical properties of tissue as indicators of disease.

Elasticity Imaging

Over the last 20 years, a number of imaging techniques have been developed to obtain information about the mechanical properties of tissue in vivo. Figure 2.2 provides a summary of the literature published on the topic of elasticity imaging over this time. This figure reflects the result of searching for "elastography" in the PubMed and Web of Knowledge databases for each publication year, ignoring false hits and conference abstracts, and differentiating the articles based on the primary imaging technique used or discussed. The imaging techniques, which will be discussed in more detail below, were crudely classified as [1] any ultrasound elasticity or strain imaging, [2] any phase-based MR elasticity or strain imaging, and [3] any other technique, including optical coherence tomography, X-ray, and MR tagging. While not an exhaustive or definitive search, the point of this figure is to indicate the rapid growth that the field of elasticity imaging has experienced, even in just the last 5 years in which the number of publications has doubled. This reflects the growing interest in the field and the rapid development of elasticity imaging techniques and applications.

Imaging the elastic properties of tissue involves several important steps. First, a static, transient, or harmonic force is applied to the tissue. Second, the resulting internal displacements of the tissue are measured. Third, the measured displacements are used to solve the equations of motion [e.g., Eq. (2.2)] that are assumed to model the behavior of the tissue to determine the mechanical properties of the tissue. Throughout the evolution of elasticity imaging, numerous approaches have been developed to

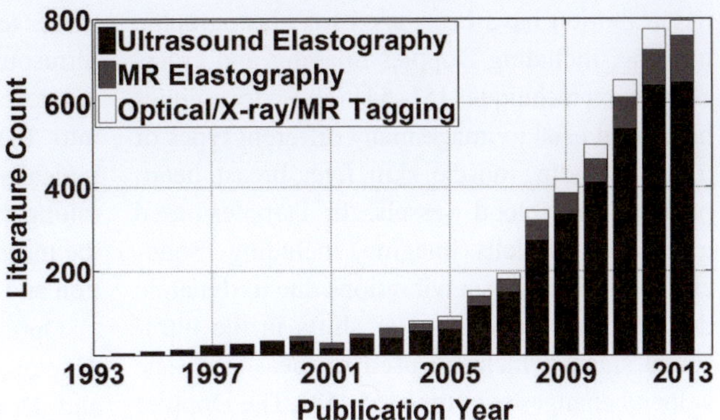

Fig. 2.2 A literature summary from searches in PubMed and Web of Knowledge using the search term "elastography." The results are roughly subdivided based on the primary imaging technique used or discussed

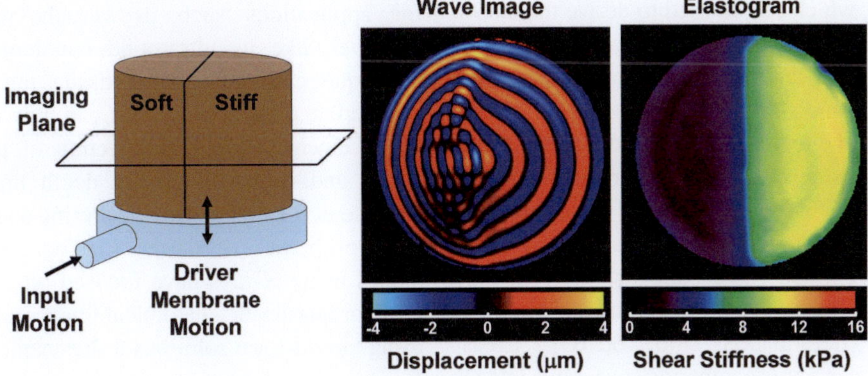

Fig. 2.3 Demonstration of elasticity imaging of a tissue-mimicking phantom. The cylindrical phantom consists of a soft half and a stiff half. The phantom is placed on top of a pneumatically powered drum that vibrates it vertically. Cross-sectional imaging is performed to measure the internal displacement field of the phantom, which provides pictures such as the one in the middle showing the through-plane component of the displacement field. The wave image shows shear waves propagating with a short wavelength through the soft material and a long wavelength in the stiff material. The wave images are used as the input into a wave equation modeling the wave propagation to solve for the stiffness of the phantom, as shown in the elastogram on the right

accomplish each of the above steps, each with its own benefits and limitations. Figure 2.3 shows a classic phantom experiment used to demonstrate these elasticity imaging principles.

The process of producing tissue motion can be performed using both extrinsic and intrinsic sources. Classic examples of extrinsic sources of motion include the use of piezoelectric elements, electromechanical actuators, and pneumatically powered drums and tubes that are placed in contact with the body [37, 38]. The vibrations produced by these external devices propagate into the body and throughout the tissue of interest.

In contrast, changes in blood and CSF pressure, as well as the motion of the beating heart, can serve as sources of intrinsic motion and can produce motion directly within tissue [39–43]. Focused-ultrasound techniques, such as using amplitude modulation or interfering ultrasound beams, have also been employed as an external source of vibration that is capable of producing motion directly within tissue [44–50].

Tissue displacement can be measured in vivo using a number of techniques, including ultrasound, MRI, X-ray, and optical methods. A number of ultrasound-based methods for measuring tissue motion have been used for elastographic imaging, including Doppler imaging and cross-correlation techniques [51–54]. These techniques have been used to image many different types of tissue, including muscle, skin, liver, breast, heart, prostate, and blood vessels. In Doppler-based ultrasound elasticity imaging, including "sono-elasticity," local tissue vibrations due to dynamic deformations cause Doppler shifts in the ultrasound signal which are used to measure tissue velocity changes over time [55–58]. The Doppler information can be used to obtain images showing wave propagation through tissue from which shear wavelengths and wave speeds can be obtained, which can be used to derive the elastic properties of the tissue [56, 59]. This record of tissue motion can also be used to calculate displacement and strain, as in the speckle-tracking techniques below, or it can be used with an appropriate equation of motion to directly solve for quantitative tissue mechanical properties. In speckle-tracking ultrasound elastography techniques, repeated ultrasound acquisitions are obtained while tissue is undergoing quasistatic or dynamic deformations [40, 60–63]. Cross-correlation analysis comparing successive images allows for estimates of the tissue displacement between the two acquisitions. The measured displacements can then be used to calculate strain maps (qualitative images of tissue elasticity) which are used to characterize the tissue [64–67] or to track wave propagation through the tissue for purposes of calculating the wave speed, which relates to tissue stiffness [62, 63]. The transient elastography method, also called Fibroscan, has become a very popular technique for quickly assessing hepatic fibrosis in vivo and noninvasively [62, 68]. Ultrasound elastography using ultrafast imaging techniques perform speckle-tracking to measure tissue displacement, but use customized hardware to acquire and reconstruct images at an effective frame rate of about 5,000 images per second [46, 63]. This method can image steady-state and transient wave propagation with high temporal and spatial resolutions and can be used to estimate tissue mechanical properties, including anisotropic and nonlinear properties [69, 70]. The primary limitations of these techniques are those produced by the ultrasound imaging itself in that an acoustic window is required to get the ultrasound signal into the tissue, the tissue displacement can typically only be measured in one direction (along the direction of the ultrasound beam), and the measurement depth is limited by the attenuation and scattering of the ultrasound beam.

Optical elastography methods have also been developed which are similar to speckle-tracking and Doppler ultrasound elastography, but use light instead of ultrasound to measure the tissue motion [71–76]. The higher resolution achievable with optical elastography makes it desirable for applications such as vascular imaging. However, like its ultrasound counterpart, this technique can only image tissue through an appropriate optical window, it can only measure displacements along the direction of the light beam, and it has significant depth limitations because of the high optical scattering and attenuation properties of tissue. Elastography techniques using X-rays have focused on acquiring static or quasistatic anatomical images and either using closed-form solutions to mechanical models of tissue deformation assuming certain geometries to determine mechanical properties, or using finite-element methods to iteratively adjust the assumed mechanical properties of a model until the model of the tissue behaves like the observed tissue deformation [77–80]. The X-ray and optical elastography techniques have not been as well developed as their ultrasound and MR counterparts.

Numerous MRI techniques have been developed over the years for measuring tissue motion and for determining the mechanical properties of tissue. Several methods have been developed which use the technique of spatial modulation of magnetization (SPAMM) to modulate the longitudinal magnetization of tissue to produce banding or grid patterns in the MR images that move with the deformation of the tissue [81–84]. From the MR images, the deformation of the grid pattern can be determined and used to estimate the strain distribution throughout the tissue. These techniques have been used for assessing cardiac and skeletal muscle function as well as for studying intrinsic brain motion and models of traumatic brain injury [82, 85–90]. Like many of the speckle-tracking ultrasound and optical techniques, this technique only provides qualitative estimates of tissue mechanical properties unless additional assumptions are made about the tissue boundary conditions or the characteristics of the applied stresses. Another limitation of SPAMM techniques is that they typically require large-amplitude motion in order for the grid pattern to be tracked. While most SPAMM-based techniques produce in-plane grid patterns to measure tissue displacement, a similar approach using through-plane SPAMM tagging was developed in which the quasistatic compression of tissue can be used to produce strain-encoded images [91]. Because it does not actually track the motion of the SPAMM tags, this method is less computationally demanding than the conventional SPAMM technique, but it still only produces qualitative images of tissue mechanical properties.

In the SPAMM-based elastographic imaging techniques, tissue motion is calculated from a series of MR magnitude images. An alternative approach is to encode tissue motion into the phase of the MR signal. Normally, the magnetic field gradients used in MRI are used to determine the position of tissue that is assumed to be static. However, motion that occurs during these gradients will result in additional phase in the measured transverse magnetization according to

$$\varphi\left(r_0,t\right) = \gamma \int_{t'=0}^{t'=t} G\left(t'\right) \cdot u\left(r_0,t'\right) dt', \quad (2.4)$$

where $\varphi(r_0, t)$ is the motion-induced phase of tissue, originally at r_0, at a time t after the creation of the transverse magnetization, γ is the proton gyromagnetic ratio, $G(t)$ is the magnetic field gradient vector, and $u(r_0, t)$ models the motion of the tissue at r_0 [92, 93]. This characteristic of the MR signal has been used for such applications as diffusion imaging, flow imaging, and compensating for flow artifacts [92–96].

Several MR elastography (MRE) techniques using phase-based motion encoding have been developed to measure the quasistatic and harmonic deformation of tissue. One of the MRE techniques developed involves the imaging of quasistatic tissue compression using a stimulated-echo MR acquisition [97–100]. Stimulated-echo MRE utilizes three RF pulses: the first RF pulse creates the transverse magnetization of the tissue while the tissue is in its initial state, the second RF pulse temporarily stores a portion of that signal as longitudinal magnetization while the tissue is deformed from its initial state (during the "mixing" time of the stimulated-echo sequence), and the third RF pulse returns the stored longitudinal magnetization to the transverse plane to be measured while the tissue is still in its deformed state. To determine the amount of motion that the tissue has experienced, motion-encoding gradient pulses are incorporated into the imaging sequence before and after the mixing time. For tissue that does not move during the mixing time, the effects of the motion-encoding gradients (MEG) cancel each other and no net phase shift is recorded. However, in regions where the tissue has moved, a phase shift will be recorded that is proportional to the amount of displacement that has occurred. Like the SPAMM technique, this record of the tissue displacement that is present in the MR images can be used to calculate the strain distribution within the tissue and to provide qualitative measures of tissue mechanical properties. To determine the tissue properties quantitatively, additional assumptions must be made about the boundary conditions and the stresses applied to the tissue.

An alternative form of MRE emerged from the desire to image the response of tissue to dynamic stresses [101, 102]. By using time-harmonic motion, equations of motion like Eq. (2.2) could be used to directly determine tissue mechanical properties, even without knowledge of the boundary conditions present during the acquisition. Early work showed that MR imaging sequences could be easily modified to be sensitive to periodic motion and that the properties of the acquisition could be tailored to enhance the sensitivity to motion at a particular frequency, or to yield broadband motion sensitivity, depending on the application [102–107]. Much like velocity encoding in MR vascular flow imaging, the motion sensitivity of MRE is accomplished by adding

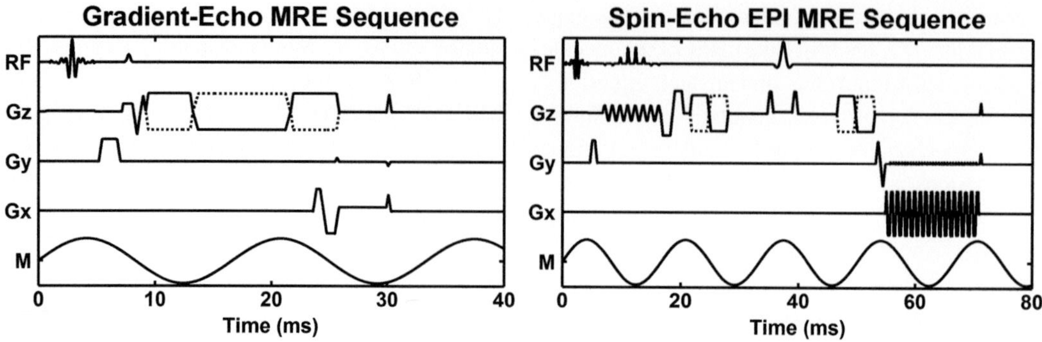

Fig. 2.4 Examples of gradient-echo MRE and spin-echo EPI MRE pulse sequences. The diagrams show the RF waveforms; the X, Y, and Z gradient waveforms, and the tissue motion ("M"). The MEG, indicated by the alternating polarity *solid* and *dashed curves* in each example, are shown applied in the Z direction to image just that component of the tissue motion. In general, the MEG can be applied along any combination of axes to record the component(s) of the motion needed for the analysis

additional MEG into standard MR imaging sequences which encode additional phase into the MR images in accordance with Eq. (2.4). Examples of MR imaging sequences which have been modified for MRE include gradient-recalled echo (GRE), spin echo (SE), echo planar imaging (EPI), balanced steady-state free precession (bSSFP), spiral, and stimulated-echo sequences [101–103, 106, 108–113]. Examples of GRE and SE-EPI MRE pulse sequences are shown in Fig. 2.4. Also like vascular flow imaging, MRE acquisitions are often repeated with two different amplitudes for the MEG (as indicated by the dashed MEG in Fig. 2.4). These two datasets are then combined to increase motion sensitivity while removing static phase errors from the images. In most applications, the MRE phase data are considered to be directly proportional to the tissue motion, thus the phase data can be substituted for displacement in equations of motion, such as Eq. (2.2), to determine the tissue mechanical properties.

Because the tissue motion in dynamic MRE is dynamic, the imaging sequence and the motion must be synchronized so that the timing, or phase relationship, between the motion and MEG remains constant during the acquisition. Each phase image produced in this fashion (also called a "wave image") is equivalent to imaging the dynamic displacement field at one instant in time.

By changing the timing between the motion and MEG from one image acquisition to the next, multiple images ("phase offsets") can be obtained which show the propagation of the displacement field over time. The time-domain data can then either be processed directly, or analyzed in the Fourier domain to isolate the motion occurring at specific frequencies. Besides continuous periodic vibrations, the propagation of transient impulses through tissue can also be imaged. This technique has been pursued as an alternative means for measuring strain, stiffness, and other tissue properties, and has also been considered as a model for studying tissue response to traumatic injury [114–116].

Calculating Mechanical Properties

In order to determine the mechanical properties of tissue, the displacement data obtained from an elastographic imaging technique must be processed using algorithms based on appropriate equations of motion that characterize the tissue response to the applied stresses used during the acquisition. For MRE, numerous algorithms have been developed incorporating variational methods, finite-element models (FEM), level set techniques, and the direct solution of the differential equations of motion [117]. From one application to another, different assumptions have to be made about the tissue, including assumptions about tissue anisotropy, compressibility, attenuation, dispersion, poroelasticity, and geometric effects (e.g., waveguide effects and wave scattering properties). The images that are obtained from these inversion algorithms, which indicate the distribution of tissue mechanical properties, are often referred to as elastograms. Using such algorithms, it is possible to estimate tissue properties such as the shear modulus, shear wave speed, and shear viscosity, as well as to produce images indicating tissue structural information such as anisotropy and the fluid–solid nature of the tissue.

For many elastography applications, tissue is assumed to be linearly elastic, homogeneous, and isotropic. These assumptions result in the approximation that the shear modulus μ of the tissue can be expressed as

$$\mu = \rho c_s^2 = \rho \left(\lambda f \right)^2, \qquad (2.5)$$

where ρ is the density of the tissue, c_s is the shear wave speed in the tissue, λ is the shear wavelength, and f is the frequency of tissue vibration. If the tissue is dispersive, the shear modulus of the tissue changes with the frequency of vibration. Therefore, the shear modulus obtained using Eq. (2.5) for a particular frequency is only valid at that frequency, and this single-frequency estimate of the shear modulus is often called the shear stiffness. One way to determine the shear modulus from Eq. (2.5) is to estimate the shear wavelength from profiles extracted from the measured displacement data. Different techniques have been demonstrated for estimating the shear modulus in this fashion, including measuring the distance between peaks and troughs of the waves [118–120], fitting a sinusoid to the displacement data [48, 121], and fitting lines to the phase of complex-valued displacement data (the so-called "phase gradient" method) [122, 123]. While effective and straightforward to implement, these methods can be time consuming and error prone in the presence of interfering waves. A more sophisticated and robust method for measuring tissue stiffness was achieved with the implementation of the local frequency estimation

(LFE) algorithm, which uses a series of image filters with different center frequencies and bandwidths to obtain automatic and isotropic estimates of stiffness [117, 124]. The multiscale nature of the LFE algorithm produces robust estimates of stiffness for heterogeneous media even in the presence of noise, making the technique appealing for a number of MRE applications, including kidney and prostate imaging [125–130]. However, the LFE technique can have lower spatial resolution than other techniques; often underestimates the stiffness of small, stiff structures; and lacks the ability to estimate the viscous properties of a material.

When the equations of motion governing tissue displacement are known, such as in Eq. (2.2), the tissue properties can sometimes be determined by directly substituting the measured displacements into the equations. This method is referred to as performing a direct inversion (DI) of the differential equations of motion [117, 131–135]. Because they are based directly on the equations of motion, DI algorithms can be designed that incorporate viscosity, anisotropy, and geometric effects [134, 136–139]. Also, since the analysis is typically performed on small, local regions of tissue, DI methods tend to have better spatial resolution than LFE methods. While DI methods are frequently applied to heterogeneous media, the inversions themselves typically assume that the mechanical properties of the medium are uniform within each processing window. This "local homogeneity" assumption is usually required for a practical solution of the equations of motion and is most valid when processing data within small regions away from boundaries and tissue interfaces. Since DI methods often require the calculation of high-order derivatives of noisy displacement data, when the data have low SNR, stiffness estimates provided by DI often have a larger variance than those produced by LFE. Careful consideration has to be made to the degree of data smoothing that is performed and the technique used to estimate the derivatives of the data to reduce the impact of noise on the final elastograms.

An alternative to directly inverting the differential equations of motion to obtain estimates of material mechanical properties, which involves directly taking high-order derivatives of noisy data, are inversions which incorporate data models or smooth "test functions," so some of the derivatives are calculated on smooth analytic functions rather than noisy measured data. These techniques include classic variational methods as well as alternative formulations of the equations of motion that require fewer derivatives of the noisy data [135, 136, 140]. Variational methods, specifically, perform integration on the data while shifting the derivative calculations to well-behaved, smooth functions introduced partly to localize the inversion. While these methods would seem to be more robust to noise by design, in practice they are often implemented in a way that is equivalent to DI methods with some smoothing of the data, and thus they have only shown modest advantages over DI techniques. In contrast to techniques like DI and variational methods, which are often designed to ignore tissue geometry and heterogeneity, inversion algorithms based on the use of FEM have been designed which directly incorporate information about the tissue geometry and the full equations of motion [141–146]. In FEM inversions, a model is created using the imaged tissue geometry and motion, the equations of motion, an initial distribution of tissue mechanical properties, and an approximation for the boundary conditions. The FEM is then solved for the displacement field throughout the tissue. This displacement field is compared to the measured displacement field and the tissue mechanical properties in the model are then adjusted and the FEM is solved again. After a number of iterations, the distribution of tissue mechanical properties in the model converges to a final estimate of the tissue properties that produce a displacement field that is most similar to the measured displacements. This technique can be very robust to noise since it does not require the calculation of derivatives of noisy data, and it can more accurately model the wave propagation in tissue because it does not make the local homogeneity assumption required by other algorithms. However, it also requires an accurate model of the tissue motion, geometry, and boundary conditions for an accurate simulation, and this information can be difficult to obtain in practice.

Also, because of the repeated evaluation of the FEM model, which may be very large for the general 3D case, these methods are often very time consuming.

The presence of interfering waves in the measured MRE displacement data can cause artifacts in some MRE inversion algorithms (e.g., the phase-gradient algorithm, DI methods, and LFE). This is because some algorithms implicitly or explicitly assume that the analysis is being performed on a single plane wave rather than a more complicated wave field with multiple interfering wave fronts, while the destructive interference that can arise from interfering waves causes numerical instabilities in other algorithms. A technique called directional filtering was developed to reduce these artifacts by using the temporal and spatial information in MRE data to separate waves propagating in different directions [147]. The extracted waves propagating in various directions each constitute a new displacement field on which the inversion algorithm can be applied to produce an estimate of tissue mechanical properties, and these multiple estimates of the mechanical properties can be averaged together to form the final elastogram. Additional low-pass and high-pass filtering properties can be incorporated into the directional filters to reduce image noise and the effects of longitudinal wave propagation and tissue bulk motion, both of which tend to introduce long wavelength information into the measured displacement data. Longitudinal wave motion can be better removed from the data by calculating the vector curl of the measured wave field, though this is not always possible as it requires the measurement of the full 3D vector wave field, which may be too time consuming in some applications [109, 148, 149]. An alternative approach to the directional filter for reducing the effects of interfering waves is to design a model of the wave propagation that incorporates waves propagating in many different directions and solves for them simultaneously [136]. Additional improvements in the visualization of tissue mechanical properties can be obtained by incorporating data measured using different mechanical vibration frequencies. The wave fields produced by different vibration frequencies have different regions of wave interference. Merging the different wave fields in one inversion compensates for this wave interference and improves the overall depiction of the tissue properties [107, 136]. One limitation of this method, however, is that it assumes that the tissue mechanical properties are the same at these different vibration frequencies. Due to dispersive effects in the tissue, this is not true. However, the difference in properties may be small enough over a small enough bandwidth of frequencies that the error in this type of tissue model is insignificant to the overall improvement in the way the tissue properties are depicted.

Summary

The field of elasticity imaging has evolved tremendously over the last 30 years and has included significant developments at every level, from improvements in how motion is created in vivo and what types of motion can be imaged, to how the imaging of tissue motion is performed, to how the images of tissue motion are processed to reveal tissue mechanical properties. In MRE, these developments have allowed for the in vivo evaluation of the liver, spleen, kidneys, brain, breast, skeletal muscle, and heart, among other applications, to assess their normal properties and how these mechanical properties change with the progression of various diseases. The rest of this book will discuss the current state of the art in MRE applications for imaging these various organs.

References

1. Field D. Anatomy: palpation and surface markings. 3rd ed. Oxford: Butterworth Heinemann; 2001.
2. Bickley LS, Szilagyi PG. Bates' guide to physical examination and history taking. 8th ed. Philadelphia: Lippincott Williams & Wilkins; 2003.
3. Mangione S. Physical diagnosis secrets. 2nd ed. Philadelphia: Mosby/Elsevier; 2008.
4. Bacon BR, O'Grady JG, Di Bisceglie AM, Lake JR. Comprehensive clinical hepatology. 2nd ed. Philadelphia: Elsevier Limited; 2006.
5. Barton MB, Harris R, Fletcher SW. Does this patient have breast cancer? The screening clinical breast examination: should it be done? How? JAMA. 1999;282(13):1270–80.
6. Roubidoux MA, Bailey JE, Wray LA, Helvie MA. Invasive cancers detected after breast cancer screening yielded a negative result: relationship of mammographic density to tumor prognostic factors. Radiology. 2004;230(1):42–8.
7. Roubidoux MA, Wilson TE, Orange RJ, Fitzgerald JT, Helvie MA, Packer SA. Breast cancer in women who undergo screening mammography: relationship of hormone replacement therapy to stage and detection method. Radiology. 1998;208(3):725–8.
8. Guimaraes CM, Correia MM, Baldisserotto M, Aires EPD, Coelho JF. Intraoperative ultrasonography of the liver in patients with abdominal tumors - a new approach. J Ultrasound Med. 2004;23(12):1549–55.
9. Fung YC. Biomechanics: mechanical properties of living tissues. 2nd ed. New York: Springer; 1993.
10. Duck FA. Physical properties of tissue: a comprehensive reference book. London: Academic; 1990.
11. Holzapfel BM, Reichert JC, Schantz JT, Gbureck U, Rackwitz L, Noth U, et al. How smart do biomaterials need to be? A translational science and clinical point of view. Adv Drug Deliv Rev. 2013;65(4):581–603.
12. Tritto G, Davies NA, Jalan R. Liver replacement therapy. Semin Respir Crit Care Med. 2012;33(1):70–9.
13. Malchesky PS. Artificial organs 2011: a year in review. Artif Organs. 2012;36(3):291–323.
14. Atala A. Engineering organs. Curr Opin Biotechnol. 2009;20(5):575–92.
15. Yang F, Li JCM. Impression test – a review. Mater Sci Eng R. 2013;74(8):233–53.
16. Walley SM. Historical origins of indentation hardness testing. Mater Sci Technol. 2012;28(9–10):1028–44.
17. Lakes RS. Viscoelastic materials. Cambridge: Cambridge University Press; 2009.
18. Nerurkar NL, Elliott DM, Mauck RL. Mechanical design criteria for intervertebral disc tissue engineering. J Biomech. 2010;43(6):1017–30.
19. Laksari K, Shafieian M, Darvish K. Constitutive model for brain tissue under finite compression. J Biomech. 2012;45(4):642–6.
20. Sparrey CJ, Keaveny TM. Compression behavior of porcine spinal cord white matter. J Biomech. 2011; 44(6):1078–82.
21. Stokes IA, Laible JP, Gardner-Morse MG, Costi JJ, Iatridis JC. Refinement of elastic, poroelastic, and osmotic tissue properties of intervertebral disks to analyze behavior in compression. Ann Biomed Eng. 2011;39(1):122–31.
22. Hatami-Marbini H, Etebu E. An experimental and theoretical analysis of unconfined compression of corneal stroma. J Biomech. 2013;46(10):1752–8.
23. Mansour JM, Welter JF. Multimodal evaluation of tissue-engineered cartilage. J Med Biol Eng. 2013; 33(1):1–16.
24. Van Loocke M, Lyons CG, Simms CK. Viscoelastic properties of passive skeletal muscle in compression: stress-relaxation behaviour and constitutive modelling. J Biomech. 2008;41(7):1555–66.

25. Chen DL, Yang PF, Lai YS. A review of three-dimensional viscoelastic models with an application to viscoelasticity characterization using nanoindentation. Microelectron Reliab. 2012;52(3):541–58.
26. O'Connor WT, Smyth A, Gilchrist MD. Animal models of traumatic brain injury: a critical evaluation. Pharmacol Ther. 2011;130(2):106–13.
27. Lin YC, Chen XM. A critical review of experimental results and constitutive descriptions for metals and alloys in hot working. Mater Des. 2011;32(4):1733–59.
28. Murphy JG. Transversely isotropic biological, soft tissue must be modelled using both anisotropic invariants. Eur J Mech-A/Solids. 2013;42:90–6.
29. Cowin SC. Bone poroelasticity. J Biomech. 1999;32(3):217–38.
30. Darvish KK, Crandall JR. Nonlinear viscoelastic effects in oscillatory shear deformation of brain tissue. Med Eng Phys. 2001;23(9):633–45.
31. Hasan A, Ragaert K, Swieszkowski W, Selimovic S, Paul A, Camci-Unal G, et al. Biomechanical properties of native and tissue engineered heart valve constructs. J Biomech. 2013;47(9):1949–63. doi:10.1016/j.jbiomech.2013.09.023 Oct 21.
32. Aki K, Richards PG. Quantitative seismology. 2nd ed. Sausalito: University Science; 2002.
33. Oestreicher HL. Field and impedance of an oscillating sphere in a viscoelastic medium with application to biophysics. J Acoust Soc Am. 1951;23(6):707–14.
34. Auld BA. Acoustic fields and waves in solids. 2nd ed. Malabar: Krieger; 1990.
35. Sarvazyan AP, Skovoroda AR, Emelianov SY, Fowlkes JB, Pipe JG, Adler RS, et al. Biophysical bases of elasticity imaging. In: Jones JP, editor. Acoustical imaging. New York: Plenum; 1995. p. 223–40.
36. Krouskop TA, Wheeler TM, Kallel F, Garra BS, Hall T. Elastic moduli of breast and prostate tissues under compression. Ultrason Imaging. 1998;20(4):260–74.
37. Uffmann K, Ladd ME. Actuation systems for MR elastography: design and applications. IEEE Eng Med Biol Mag. 2008;27(3):28–34.
38. Tse ZT, Janssen H, Hamed A, Ristic M, Young I, Lamperth M. Magnetic resonance elastography hardware design: a survey. Proc Inst Mech Eng H. 2009;223(4):497–514.
39. Dighe M, Bae U, Richardson ML, Dubinsky TJ, Minoshima S, Kim Y. Differential diagnosis of thyroid nodules with US elastography using carotid artery pulsation. Radiology. 2008;248(2):662–9.
40. Konofagou EE, D'Hooge J, Ophir J. Myocardial elastography – a feasibility study in vivo. Ultrasound Med Biol. 2002;28(4):475–82.
41. Hirsch S, Klatt D, Freimann F, Scheel M, Braun J, Sack I. In vivo measurement of volumetric strain in the human brain induced by arterial pulsation and harmonic waves. Magn Reson Med. 2013;70(3):671–83.
42. Kolen AF, Miller NR, Ahmed EE, Bamber JC. Characterization of cardiovascular liver motion for the eventual application of elasticity imaging to the liver in vivo. Phys Med Biol. 2004;49(18):4187–206.
43. Chung S, Breton E, Mannelli L, Axel L. Liver stiffness assessment by tagged MRI of cardiac-induced liver motion. Magn Reson Med. 2011;65(4):949–55.
44. Nightingale K, Soo MS, Nightingale R, Trahey G. Acoustic radiation force impulse imaging: in vivo demonstration of clinical feasibility. Ultrasound Med Biol. 2002;28(2):227–35.
45. Sarvazyan AP, Rudenko OV, Swanson SD, Fowlkes JB, Emelianov SY. Shear wave elasticity imaging: a new ultrasonic technology of medical diagnostics. Ultrasound Med Biol. 1998;24(9):1419–35.
46. Sinkus R, Tanter M, Bercoff J, Siegmann K, Pernot M, Athanasiou A, et al. Potential of MRI and ultrasound radiation force in elastography: applications to diagnosis and therapy. Proc IEEE. 2008;96(3):490–9.
47. McDannold N, Maier SE. Magnetic resonance acoustic radiation force imaging. Med Phys. 2008;35(8):3748–58.
48. Wu T, Felmlee JP, Greenleaf JF, Riederer SJ, Ehman RL. MR imaging of shear waves generated by focused ultrasound. Magn Reson Med. 2000;43(1):111–5.
49. Greenleaf JF, Fatemi M, Insana M. Selected methods for imaging elastic properties of biological tissues. Annu Rev Biomed Eng. 2003;5:57–78.
50. Chen S, Urban MW, Pislaru C, Kinnick R, Zheng Y, Yao A, et al. Shearwave dispersion ultrasound vibrometry (SDUV) for measuring tissue elasticity and viscosity. IEEE Trans Ultrason Ferroelectr Freq Control. 2009;56(1):55–62.
51. Parker KJ, Doyley MM, Rubens DJ. Imaging the elastic properties of tissue: the 20 year perspective. Phys Med Biol. 2011;56:R1–29.
52. Wells PN, Liang HD. Medical ultrasound: imaging of soft tissue strain and elasticity. J R Soc Interface. 2011;8(64):1521–49.
53. Sporea I, Gilja OH, Bota S, Sirli R, Popescu A. Liver elastography – an update. Med Ultrason. 2013;15(4):304–14.
54. Castaneda B, Ormachea J, Rodriguez P, Parker KJ. Application of numerical methods to elasticity imaging. Mol Cell Biomech: MCB. 2013;10(1):43–65.
55. Krouskop TA, Dougherty DR, Vinson FS. A pulsed Doppler ultrasonic system for making noninvasive measurements of the mechanical properties of soft tissue. J Rehabil Res Dev. 1987;24(2):1–8.
56. Yamakoshi Y, Sato J, Sato T. Ultrasonic imaging of internal vibration of soft tissue under forced vibration. IEEE Trans Ultrason Ferroelectr Freq Control. 1990;37(2):45–53.
57. Lerner RM, Huang SR, Parker KJ. "Sonoelasticity" images derived from ultrasound signals in mechanically vibrated tissues. Ultrasound Med Biol. 1990;16(3):231–9.
58. Parker KJ, Huang SR, Musulin RA, Lerner RM. Tissue response to mechanical vibrations for "sonoelasticity imaging". Ultrasound Med Biol. 1990;16(3):241–6.
59. Sanada M, Ebara M, Fukuda H, Yoshikawa M, Sugiura N, Saisho H, et al. Clinical evaluation of sonoelasticity measurement in liver using ultrasonic

imaging of internal forced low-frequency vibration. Ultrasound Med Biol. 2000;26(9):1455–60.

60. Ophir J, Cespedes I, Ponnekanti H, Yazdi Y, Li X. Elastography: a quantitative method for imaging the elasticity of biological tissues. Ultrason Imaging. 1991;13(2):111–34.

61. Ophir J, Kallel F, Varghese T, Konofagou E, Alam SK, Krouskop T, et al. Elastography. C R Acad Sci Ser IV-Phys Astr. 2001;2(8):1193–212.

62. Sandrin L, Fourquet B, Hasquenoph JM, Yon S, Fournier C, Mal F, et al. Transient elastography: a new noninvasive method for assessment of hepatic fibrosis. Ultrasound Med Biol. 2003;29(12):1705–13.

63. Sandrin L, Tanter M, Catheline S, Fink M. Shear modulus imaging with 2-D transient elastography. IEEE Trans Ultrason Ferroelectr Freq Control. 2002;49(4):426–35.

64. Righetti R, Righetti M, Ophir J, Krouskop TA. The feasibility of estimating and imaging the mechanical behavior of poroelastic materials using axial strain elastography. Phys Med Biol. 2007;52(11):3241–59.

65. Ophir J, Srinivasan S, Righetti R, Thittai A. Elastography: a decade of progress (2000–2010). Curr Med Imaging Rev. 2011;7(4):292–312.

66. Cespedes I, Ophir J, Ponnekanti H, Maklad N. Elastography: elasticity imaging using ultrasound with application to muscle and breast in vivo. Ultrason Imaging. 1993;15(2):73–88.

67. Ophir J, Alam SK, Garra BS, Kallel F, Konofagou EE, Krouskop T, et al. Elastography: imaging the elastic properties of soft tissues with ultrasound. J Med Ultrason. 2002;29:155–71.

68. Jung KS, Kim SU. Clinical applications of transient elastography. Clin Mol Hepatol. 2012;18(2):163–73.

69. Gennisson JL, Grenier N, Combe C, Tanter M. Supersonic shear wave elastography of in vivo pig kidney: influence of blood pressure, urinary pressure and tissue anisotropy. Ultrasound Med Biol. 2012;38(9):1559–67.

70. Sinkus R, Bercoff J, Tanter M, Gennisson JL, El Khoury C, Servois V, et al. Nonlinear viscoelastic properties of tissue assessed by ultrasound. IEEE Trans Ultrason Ferroelectr Freq Control. 2006; 53(11):2009–18.

71. Kirkpatrick SJ, Wang RK, Duncan DD, Kulesz-Martin M, Lee K. Imaging the mechanical stiffness of skin lesions by in vivo acousto-optical elastography. Opt Express. 2006;14(21):9770–9.

72. Wang RKK, Ma ZH, Kirkpatrick SJ. Tissue Doppler optical coherence elastography for real time strain rate and strain mapping of soft tissue. Appl Phys Lett. 2006;89(14):144103.

73. Rogowska J, Patel N, Plummer S, Brezinski ME. Quantitative optical coherence tomographic elastography: method for assessing arterial mechanical properties. Br J Radiol. 2006;79(945):707–11.

74. Kennedy KM, McLaughlin RA, Kennedy BF, Tien A, Latham B, Saunders CM, et al. Needle optical coherence elastography for the measurement of microscale mechanical contrast deep within human breast tissues. J Biomed Opt. 2013;18(12):121510.

75. Wang S, Larin KV. Shear wave imaging optical coherence tomography (SWI-OCT) for ocular tissue biomechanics. Opt Lett. 2014;39(1):41–4.

76. Nahas A, Tanter M, Nguyen TM, Chassot JM, Fink M, Claude BA. From supersonic shear wave imaging to full-field optical coherence shear wave elastography. J Biomed Opt. 2013;18(12):121514.

77. Lee MK, Drangova M, Holdsworth DW, Fenster A. Application of dynamic computed tomography for measurements of local aortic elastic modulus. Med Biol Eng Comput. 1999;37(1):13–24.

78. Wang ZG, Liu Y, Wang G, Sun LZ. Nonlinear elasto-mammography for characterization of breast tissue properties. Int J Biomed Imaging. 2011;2011:10. 540820.

79. Wang ZG, Liu Y, Sun LZ, Wang G, Fajardo LL. Elasto-mammography: theory, algorithm, and phantom study. Int J Biomed Imaging. 2006;2006:53050.

80. Drangova M, Holdsworth DW, Boyd CJ, Dunmore PJ, Roach MR, Fenster A. Elasticity and geometry measurements of vascular specimens using a high-resolution laboratory CT scanner. Physiol Meas. 1993;14(3):277–90.

81. Axel L, Dougherty L. Heart wall motion: improved method of spatial modulation of magnetization for MR imaging. Radiology. 1989;172(2):349–50.

82. Osman NF, Kerwin WS, McVeigh ER, Prince JL. Cardiac motion tracking using CINE harmonic phase (HARP) magnetic resonance imaging. Magn Reson Med. 1999;42(6):1048–60.

83. Wang H, Amini AA. Cardiac motion and deformation recovery from MRI: a review. IEEE Trans Med Imaging. 2012;31(2):487–503.

84. Creswell LL, Moulton MJ, Wyers SG, Pirolo JS, Fishman DS, Perman WH, et al. An experimental method for evaluating constitutive models of myocardium in in vivo hearts. Am J Physiol. 1994;267 (2 Pt 2):H853–63.

85. Yeon SB, Reichek N, Tallant BA, Lima JA, Calhoun LP, Clark NR, et al. Validation of in vivo myocardial strain measurement by magnetic resonance tagging with sonomicrometry. J Am Coll Cardiol. 2001; 38(2):555–61.

86. Gotte MJ, Germans T, Russel IK, Zwanenburg JJ, Marcus JT, van Rossum AC, et al. Myocardial strain and torsion quantified by cardiovascular magnetic resonance tissue tagging: studies in normal and impaired left ventricular function. J Am Coll Cardiol. 2006;48(10):2002–11.

87. Sabet AA, Christoforou E, Zatlin B, Genin GM, Bayly PV. Deformation of the human brain induced by mild angular head acceleration. J Biomech. 2008;41(2):307–15.

88. Soellinger M, Ryf S, Boesiger P, Kozerke S. Assessment of human brain motion using CSPAMM. J Magn Reson Imaging. 2007;25(4):709–14.

89. Fu YB, Chui CK, Teo CL, Kobayashi E. Motion tracking and strain map computation for quasi-static magnetic resonance elastography. Med Image Comput Comput Assist Interv. 2011;14(Pt 1): 428–35.

90. Ceelen KK, Stekelenburg A, Mulders JL, Strijkers GJ, Baaijens FP, Nicolay K, et al. Validation of a numerical model of skeletal muscle compression with MR tagging: a contribution to pressure ulcer research. J Biomech Eng. 2008;130(6):061015.

91. Osman NF. Detecting stiff masses using strain-encoded (SENC) imaging. Magn Reson Med. 2003;49(3):605–8.

92. Bernstein MA, King KF, Zhou ZJ. Handbook of MRI pulse sequences. Burlington: Elsevier Academic; 2004.

93. Haacke EM. Magnetic resonance imaging: physical principles and sequence design. New York: Wiley; 1999.

94. Singer JR. NMR diffusion and flow measurements and an introduction to spin phase graphing. J Phys E. 1978;11:281–91.

95. Dumoulin CL, Hart Jr HR. Magnetic resonance angiography. Radiology. 1986;161(3):717–20.

96. Nayler GL, Firmin DN, Longmore DB. Blood flow imaging by cine magnetic resonance. J Comput Assist Tomogr. 1986;10(5):715–22.

97. Chenevert TL, Skovoroda AR, O'Donnell M, Emelianov SY. Elasticity reconstructive imaging by means of stimulated echo MRI. Magn Reson Med. 1998;39(3):482–90.

98. Plewes DB, Bishop J, Samani A, Sciarretta J. Visualization and quantification of breast cancer biomechanical properties with magnetic resonance elastography. Phys Med Biol. 2000;45(6):1591–610.

99. Hardy PA, Ridler AC, Chiarot CB, Plewes DB, Henkelman RM. Imaging articular cartilage under compression-cartilage elastography. Magn Reson Med. 2005;53(5):1065–73.

100. Siegler P, Jenne JW, Boese JM, Huber PE, Schad LR. STEAM-sequence with multi-echo-readout for static magnetic resonance elastography. Z Med Phys. 2007;17(2):118–26.

101. Muthupillai R, Lomas DJ, Rossman PJ, Greenleaf JF, Manduca A, Ehman RL. Magnetic resonance elastography by direct visualization of propagating acoustic strain waves. Science. 1995;269(5232):1854–7.

102. Muthupillai R, Rossman PJ, Lomas DJ, Greenleaf JF, Riederer SJ, Ehman RL. Magnetic resonance imaging of transverse acoustic strain waves. Magn Reson Med. 1996;36(2):266–74.

103. Rump J, Klatt D, Braun J, Warmuth C, Sack I. Fractional encoding of harmonic motions in MR elastography. Magn Reson Med. 2007;57(2):388–95.

104. Lewa CJ. Magnetic resonance imaging in the presence of mechanical waves: NMR frequency modulation, mechanical waves as NMR factor, local temperature variations. Spectrosc Lett. 1991;24(1):55–67.

105. Lewa CJ, de Certaines JD. MR imaging of viscoelastic properties. J Magn Reson Imaging. 1995;5(2):242–4.

106. Garteiser P, Sahebjavaher RS, Ter Beek LC, Salcudean S, Vilgrain V, Van Beers BE, et al. Rapid acquisition of multifrequency, multislice and multidirectional MR elastography data with a fractionally encoded gradient echo sequence. NMR Biomed. 2013;26(10):1326–35.

107. Hirsch S, Guo J, Reiter R, Papazoglou S, Kroencke T, Braun J, et al. MR elastography of the liver and the spleen using a piezoelectric driver, single-shot wave-field acquisition, and multifrequency dual parameter reconstruction. Magn Reson Med. 2014; 71(1):267–77.

108. Glaser KJ, Felmlee JP, Ehman RL. Rapid MR elastography using selective excitations. Magn Reson Med. 2006;55(6):1381–9.

109. Sinkus R, Tanter M, Catheline S, Lorenzen J, Kuhl C, Sondermann E, et al. Imaging anisotropic and viscous properties of breast tissue by magnetic resonance-elastography. Magn Reson Med. 2005; 53(2):372–87.

110. Maderwald S, Uffmann K, Galban CJ, de Greiff A, Ladd ME. Accelerating MR elastography: a multi-echo phase-contrast gradient-echo sequence. J Magn Reson Imaging. 2006;23(5):774–80.

111. Bieri O, Maderwald S, Ladd ME, Scheffler K. Balanced alternating steady-state elastography. Magn Reson Med. 2006;55(2):233–41.

112. Huwart L, Salameh N, Ter Beek L, Vicaut E, Peeters F, Sinkus R, et al. MR elastography of liver fibrosis: preliminary results comparing spin-echo and echo-planar imaging. Eur Radiol. 2008;18(11):2535–41.

113. Johnson CL, McGarry MD, Van Houten EE, Weaver JB, Paulsen KD, Sutton BP, et al. Magnetic resonance elastography of the brain using multishot spiral readouts with self-navigated motion correction. Magn Reson Med. 2013;70(2):404–12.

114. McCracken PJ, Manduca A, Felmlee J, Ehman RL. Mechanical transient-based magnetic resonance elastography. Magn Reson Med. 2005;53(3): 628–39.

115. Gallichan D, Robson MD, Bartsch A, Miller KL. TREMR: table-resonance elastography with MR. Magn Reson Med. 2009;62(3):815–21.

116. Souchon R, Salomir R, Beuf O, Milot L, Grenier D, Lyonnet D, et al. Transient MR elastography (t-MRE) using ultrasound radiation force: theory, safety, and initial experiments in vitro. Magn Reson Med. 2008;60(4):871–81.

117. Manduca A, Oliphant TE, Dresner MA, Mahowald JL, Kruse SA, Amromin E, et al. Magnetic resonance elastography: non-invasive mapping of tissue elasticity. Med Image Anal. 2001;5(4):237–54.

118. Leclerc GE, Charleux F, Robert L, Ho Ba Tho MC, Rhein C, Latrive JP, et al. Analysis of liver viscosity behavior as a function of multifrequency magnetic resonance elastography (MMRE) postprocessing. J Magn Reson Imaging. 2013;38(2):422–8.

119. McCullough MB, Domire ZJ, Reed AM, Amin S, Ytterberg SR, Chen Q, et al. Evaluation of muscles affected by myositis using magnetic resonance elastography. Muscle Nerve. 2011;43(4):585–90.

120. Xu L, Chen J, Yin M, Glaser KJ, Chen Q, Woodrum DA, et al. Assessment of stiffness changes in the

ex vivo porcine aortic wall using magnetic resonance elastography. Magn Reson Imaging. 2012;30(1): 122–7.

121. Dresner MA, Rose GH, Rossman PJ, Muthupillai R, Manduca A, Ehman RL. Magnetic resonance elastography of skeletal muscle. J Magn Reson Imaging. 2001;13(2):269–76.

122. Yin M, Woollard J, Wang XF, Torres VE, Harris PC, Ward CJ, et al. Quantitative assessment of hepatic fibrosis in an animal model with magnetic resonance elastography. Magn Reson Med. 2007;58(2): 346–53.

123. Kolipaka A, Aggarwal SR, McGee KP, Anavekar N, Manduca A, Ehman RL, et al. Magnetic resonance elastography as a method to estimate myocardial contractility. J Magn Reson Imaging. 2012; 36(1):120–7.

124. Knutsson H, Westin CF, Granlund G. Local multi-scale frequency and bandwidth estimation. Proc IEEE Int Conf Image Process. 1994;1:36–40.

125. Bensamoun SF, Robert L, Leclerc GE, Debernard L, Charleux F. Stiffness imaging of the kidney and adjacent abdominal tissues measured simultaneously using magnetic resonance elastography. Clin Imaging. 2011;35(4):284–7.

126. Rouviere O, Souchon R, Pagnoux G, Menager JM, Chapelon JY. Magnetic resonance elastography of the kidneys: feasibility and reproducibility in young healthy adults. J Magn Reson Imaging. 2011; 34(4):880–6.

127. Braun J, Buntkowsky G, Bernarding J, Tolxdorff T, Sack I. Simulation and analysis of magnetic resonance elastography wave images using coupled harmonic oscillators and Gaussian local frequency estimation. Magn Reson Imaging. 2001;19(5): 703–13.

128. Clayton EH, Okamoto RJ, Bayly PV. Mechanical properties of viscoelastic media by local frequency estimation of divergence-free wave fields. J Biomech Eng. 2013;135(2):021025.

129. Sahebjavaher RS, Baghani A, Honarvar M, Sinkus R, Salcudean SE. Transperineal prostate MR elastography: initial in vivo results. Magn Reson Med. 2013; 69(2):411–20.

130. Li BN, Chui CK, Ong SH, Numano T, Washio T, Homma K, et al. Modeling shear modulus distribution in magnetic resonance elastography with piecewise constant level sets. Magn Reson Imaging. 2012;30(3):390–401.

131. Oliphant TE, Manduca A, Ehman RL, Greenleaf JF. Complex-valued stiffness reconstruction for magnetic resonance elastography by algebraic inversion of the differential equation. Magn Reson Med. 2001;45(2):299–310.

132. Papazoglou S, Hirsch S, Braun J, Sack I. Multifrequency inversion in magnetic resonance elastography. Phys Med Biol. 2012;57(8):2329–46.

133. Boulet T, Kelso ML, Othman SF. Microscopic magnetic resonance elastography of traumatic brain injury model. J Neurosci Methods. 2011;201(2): 296–306.

134. Doyley MM. Model-based elastography: a survey of approaches to the inverse elasticity problem. Phys Med Biol. 2012;57(3):R35–73.

135. Romano AJ, Bucaro JA, Ehman RL, Shirron JJ. Evaluation of a material parameter extraction algorithm using MRI-based displacement measurement. IEEE Trans Ultrason Ferroelectr Freq Control. 2000;47(6):1575–81.

136. Baghani A, Salcudean S, Honarvar M, Sahebjavaher RS, Rohling R, Sinkus R. Travelling wave expansion: a model fitting approach to the inverse problem of elasticity reconstruction. IEEE Trans Med Imaging. 2011;30(8):1555–65.

137. Romano A, Scheel M, Hirsch S, Braun J, Sack I. In vivo waveguide elastography of white matter tracts in the human brain. Magn Reson Med. 2012; 68(5):1410–22.

138. Kolipaka A, McGee KP, Araoz PA, Glaser KJ, Manduca A, Romano AJ, et al. MR elastography as a method for the assessment of myocardial stiffness: comparison with an established pressure–volume model in a left ventricular model of the heart. Magn Reson Med. 2009;62(1):135–40.

139. McLaughlin JR, Zhang N, Manduca A. Calculating tissue shear modulus and pressure by 2D log-elastographic methods. Inverse Prob. 2010;26(8): 085007.

140. Kwon OI, Park C, Nam HS, Woo EJ, Seo JK, Glaser KJ, et al. Shear modulus decomposition algorithm in magnetic resonance elastography. IEEE Trans Med Imaging. 2009;28(10):1526–33.

141. Perreard IM, Pattison AJ, Doyley M, McGarry MD, Barani Z, Van Houten EE, et al. Effects of frequency- and direction-dependent elastic materials on linearly elastic MRE image reconstructions. Phys Med Biol. 2010;55(22):6801–15.

142. Litwiller DV, Lee SJ, Kolipaka A, Mariappan YK, Glaser KJ, Pulido JS, et al. MR elastography of the ex vivo bovine globe. J Magn Reson Imaging. 2010;32(1):44–51.

143. Van Houten EEW, Miga MI, Weaver JB, Kennedy FE, Paulsen KD. Three-dimensional subzone-based reconstruction algorithm for MR elastography. Magn Reson Med. 2001;45(5):827–37.

144. McGarry MD, Van Houten EE, Johnson CL, Georgiadis JG, Sutton BP, Weaver JB, et al. Multiresolution MR elastography using nonlinear inversion. Med Phys. 2012;39(10):6388–96.

145. Perrinez PR, Kennedy FE, Van Houten EE, Weaver JB, Paulsen KD. Modeling of soft poroelastic tissue in time-harmonic MR elastography. IEEE Trans Biomed Eng. 2009;56(3):598–608.

146. Perrinez PR, Kennedy FE, Van Houten EE, Weaver JB, Paulsen KD. Magnetic resonance poroelastography: an algorithm for estimating the mechanical properties of fluid-saturated soft tissues. IEEE Trans Med Imaging. 2010;29(3):746–55.

147. Manduca A, Lake DS, Kruse SA, Ehman RL. Spatio-temporal directional filtering for improved inversion of MR elastography images. Med Image Anal. 2003;7(4):465–73.

148. Honarvar M, Sahebjavaher R, Sinkus R, Rohling R, Salcudean S. Curl-based finite element reconstruction of the shear modulus without assuming local homogeneity: time harmonic case. IEEE Trans Med Imaging. 2013;32(12):2189–99.

149. Baghani A, Salcudean S, Rohling R. Theoretical limitations of the elastic wave equation inversion for tissue elastography. J Acoust Soc Am. 2009;126(3):1541.

Liver Magnetic Resonance Elastography Technique

Meng Yin, Armando Manduca, and Roger C. Grimm

Introduction

The most important clinical application of MRE is in the evaluation of chronic liver diseases. Currently MRE is the most accurate technique for detection and staging of liver fibrosis noninvasively. Chronic liver disease is a worldwide problem, which has a variety of causes including viral, autoimmune, drug-induced, cholestatic, metabolic, and inherited disease [1]. In response to inflammation or direct toxic insult to the liver, zonal cellular injury forms and eventually leads to progressive hepatic fibrosis, which is scar tissue that develops progressively by an excessive production of extracellular matrix proteins [2]. The end result of untreated chronic liver disease is fibrosis and if progressive cirrhosis. Cirrhosis and its associated complications carry a high risk of mortality. Increasing evidence has demonstrated that, unlike cirrhosis, the early stages of fibrosis are treatable and reversible if appropriate anti-fibrotic treatment is given [3, 4]. Liver biopsy, the current gold standard for detection

and staging of liver fibrosis is limited by its invasive nature, mild but non-negligible risk of life threatening complications, sampling errors, and inter-observer variability. As anti-fibrotic therapies evolve, a reliable, noninvasive and cost-effective technique for diagnosis and follow-up assessment of hepatic fibrosis is needed to manage patients with chronic liver disease. Therefore, there is a need for a reliable, noninvasive method to assess liver fibrosis, not only to detect and stage the disease itself, but also to monitor treatment efficacy and optimize dosing.

A number of noninvasive imaging techniques have been evaluated for possible use in diagnosing hepatic fibrosis. Dynamic, contrast-enhanced MR imaging of the liver can show intensity and texture patterns that are associated with hepatic fibrosis, but these techniques are qualitative and may not be sufficiently sensitive in earlier stages of the disease [5], when it may be most advantageous to institute therapy. An early study of magnetization transfer contrast MR imaging techniques has shown certain changes in the presence of hepatic fibrosis, but the specificity may be limited. Studies of diffusion-weighted MR imaging techniques have shown reduced apparent diffusion coefficient values that are thought to represent the hydration and metabolic status of fibrotic liver, but the differences are relatively small. None of these imaging techniques have been shown to be sufficiently sensitive or specific for use as a clinical tool to detect occult yet significant hepatic fibrosis.

M. Yin, Ph.D. (✉) • A. Manduca, Ph.D.
R.C. Grimm
Department of Radiology, Mayo Clinic College of Medicine, Mayo Clinic, 200 First Street SW, Rochester, MN, USA
e-mail: yin.meng@mayo.edu;
Manduca.armando@mayo.edu;
Grimm.Roger@mayo.edu

S.K. Venkatesh and R.L. Ehman (eds.), *Magnetic Resonance Elastography*,
DOI 10.1007/978-1-4939-1575-0_3, © Springer Science+Business Media New York 2014

Elastography Techniques for Hepatic Fibrosis Detection (Transient Elastography vs. MRE)

The mechanical properties of liver tissue are strongly correlated with the extent of fibrosis, and perhaps with increased hepatic vascular resistance and the associated increase of portal venous pressure [6]. With recent advances in biomedical imaging, the ability to quantify in vivo stiffness by detecting wave propagation velocity through human tissue (elastography) using noninvasive techniques like ultrasound and MRI has become possible. At present, there are two well-developed noninvasive approaches to assess the mechanical properties of abdominal tissues in vivo. One is Transient Elastography (TE) [7, 8], which can provide localized quantitative shear stiffness based on 1-D measurements with ultrasound. The other is MR Elastography (MRE), which is an MR-based method that uses a modified phase-contrast MRI acquisition to determine the tissue stiffness distribution throughout a 2-D or 3-D region of interest [9]. Direct comparisons have been performed between TE and MRE stiffness measurements in phantoms/patients and have confirmed that there are no systematic differences between these measurements [10–13]. Transient elastography marketed as FibroScan® (Echosens, Paris, France) is an established technique for assessing hepatic fibrosis by measuring liver stiffness [14]. Many trials of TE for assessing hepatic fibrosis in vivo have been promising, in which liver stiffness appears to be a feasible and valid way to predict the severity of hepatic fibrosis for selected patients with advanced histological disease [15–20]. However, since the region assessed with TE encompasses a 1×4 cm cylindrical area (approximately 1/100th of the entire hepatic parenchyma) with a measurement depth between 25 and 65 mm below the skin, and is confined to the peripheral part of the right liver lobe, there is a potential for sampling errors. In the presence of prominent central obesity (defined as a body mass index > 28 kg/m^2), there is a nine-fold increased risk of technical failure with TE. Compared to ultrasound-based techniques for elasticity imaging, MR-based techniques offer some significant advantages including (1) a freely adjustable field of view, (2) no need for an acoustical window, (3) less sensitivity to body habitus such as obesity and ascites or bowel interposition, (4) the ability to obtain a conventional MR examination at the same time compared to the dedicated device used for TE, and (5) the potential to assess the entire abdomen with rapid multislice or 3-D imaging methods. Over the years, MRE has been developed for many applications to quantitatively assess the viscoelastic properties of various soft tissues, including brain [21–23], breast [24–27], heart [28, 29], lung [30], muscle [31], liver [5], spleen [32], kidneys [33], and pancreas [34]. MRE is well tolerated by most patients and has been incorporated into a standard liver MRI study. Preliminary studies have shown that MRE is feasible for the assessment of most abdominal organs [35]. Technical innovations continue to evolve in improving MRE as a technique for clinical applications. The most commonly used technique is two-dimensional (2D) gradient-recalled echo (GRE) based MRE. Since the base sequence is GRE based it is susceptible to signal loss due to presence of paramagnetic substances such as iron in patients with liver iron overload conditions such as hemochromatosis, thalassaemia etc. Alternative sequences with modifications to overcome these limitations have been developed [36]. Alternatively MRE may be performed in these patients when liver iron content returns to normal, for example after phlebotomies in the case of hemochromatosis.

Liver MRE Technique

The practical in vivo hepatic MRE technique for clinical investigations in humans is the culmination of technology development and optimization involving driver design (generating shear waves in the liver), pulse sequence design, imaging parameters, and inversion algorithms. The liver MRE technique has been well documented and can be implemented on a conventional MRI system with minimal modifications to the hardware and software. Several techniques for performing liver MRE exist. Common to all the techniques is the principle of producing mechanical shear waves in the liver with the use of a driver applied closely to the liver and then imaging the propagation of these shear waves through the liver using modified MRI sequences. Information thus obtained with the sequence is used to generate the stiffness map of the liver.

2D GRE MRE is the most common technique used for performing liver MRE and available at several leading institutions around the world [35, 37, 38].

Hardware and Setup

The hardware consists of an active driver, a passive driver, and a plastic pipe that connects the active driver to the passive driver (Fig. 3.1). The acoustic driving system (active driver) is used to generate shear waves in tissue, typically at frequencies between 40 and 200 Hz. Acoustic waves have been created in targeted tissues using active electromechanical voice coils [39, 40, 41], passive rigid rod drivers [42, 43], and passive pneumatic drum drivers [32, 44–46]. There have

also been a number of other application-specific drivers, such as a needle-based acoustic driver system used to study the liver tissue of small animal models in vivo [47]. The passive driver used in several studies involves a disc like, non-metallic passive drum driver 19 cm in diameter with a membrane that vibrates. It is activated with varying acoustic pressure conducted via a 7.6-m long plastic tube (PVC, polyvinyl chloride) from the active driver located outside the scanner room. Among the aforementioned designs, the passive pneumatic driver has several advantages. It can be easily applied against the body (such as against the thoracic and abdominal walls) in any orientation, even adjacent to the imaging coils, while avoiding problems with artifacts and heat buildup typical of electromechanical drivers.

As illustrated in Fig. 3.1, patients are usually imaged in the supine position with the passive driver placed against the anterior body wall over the right lobe of the liver on the chest wall below the breast, held in place with an elastic band around the body. The placement of the passive driver can be standardized by using xiphisternum as the landmark for placing the center of the driver and mid clavicular line for placing over the

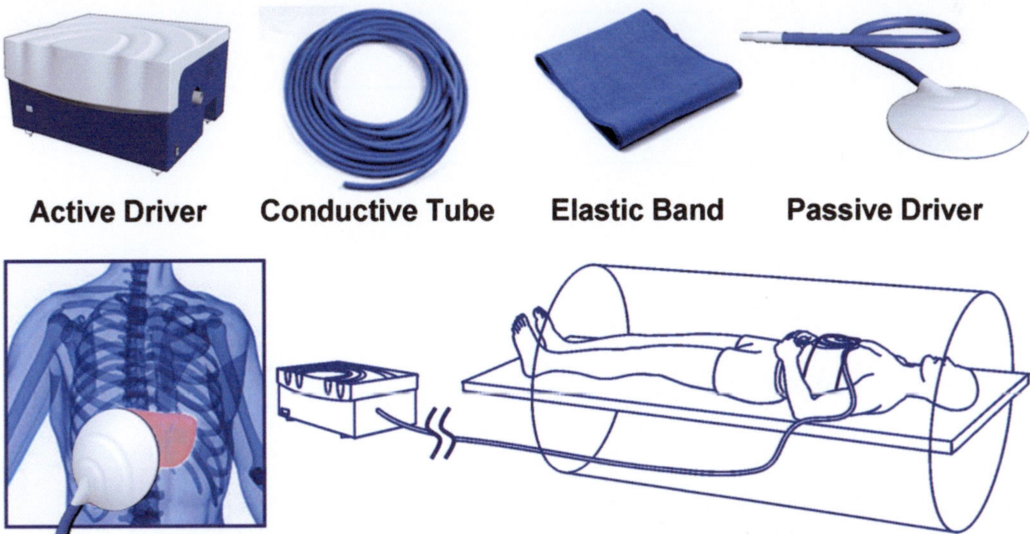

Fig. 3.1 System for applying shear waves to the abdomen for MR Elastography of the liver. All the add-on hardware elements are listed in the *top row*. Acoustic pressure waves (at 60 Hz) are generated by an active driver, located away from the magnetic field of the MRI unit, and transmitted via a flexible conductive tube to a passive driver placed over the anterior body wall. The *left bottom* diagram illustrates the location of the passive driver with respect to the liver in a coronal sketch. The *right bottom* diagram is an overview of the liver MRE examination setup

right lobe of the liver. Continuous longitudinal vibrations at 60 Hz are generated by varying acoustic pressure waves transmitted from an active driver device via a PVC conductive tube.

Software (Pulse Sequence)

The measurement of motion with MR in medical applications has been motivated significantly by the desire for blood flow quantification and vascular system imaging [48]. In the 1990s, measurement of tissue mechanical properties using MR methods was demonstrated [9, 49]. MRE uses a modified phase-contrast imaging sequence to detect propagating shear waves within the tissue of interest. Depending on the specific application, the MRE sequence uses a conventional MR imaging sequence [e.g., GRE, spin echo (SE), balanced steady-state free precession (bSSFP), or echo planar imaging (EPI)] with the inclusion of additional motion-encoding gradients (MEGs), which allow shear waves with amplitudes in the micron range to be readily imaged [9, 49–55]. The MEGs are imposed along

a specific direction and switched in polarity at an adjustable frequency. Trigger pulses synchronize the imaging sequence with the driving system to induce shear waves in the tissue, usually at the same frequency as the MEGs. In the received MR signal, a measurable phase shift caused by the cyclic motion of the spins in the presence of these MEGs can be used to calculate the displacement at each voxel, which provides a snapshot of the mechanical waves propagating within the tissue. By adjusting the phase offset between the mechanical excitation and the oscillating MEGs, wave images can be obtained for various time points of the vibration. Four phase offsets evenly spaced over 1 cycle of the motion are obtained in most experiments. This allows for the extraction of the harmonic component of the phase data at the driving frequency, giving the amplitude and the phase of the harmonic displacement at each point in space, which is the input to most MRE inversion algorithms.

A typical 2-D GRE MRE sequence (as shown in Fig. 3.2) has the following parameters (more in Table 3.1): 60 Hz continuous sinusoidal vibration, FOV = 30-42 cm, matrix = 256×64, flip

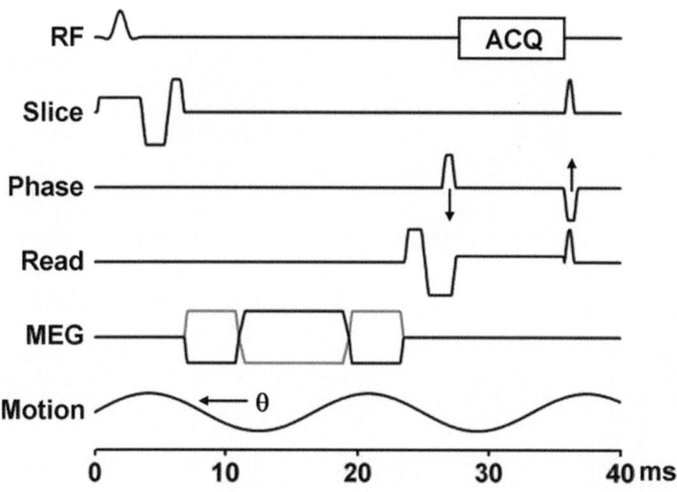

Fig. 3.2 2D GRE MRE sequence. This sequence is a conventional gradient echo sequence with an additional motion-encoding gradient (MEG) applied along the slice selection direction to detect cyclic motion in the through-plane direction. Sequence provides options to select other motion-encoding directions. The MEG is designed to minimize zeroth and first gradient moments. The MEG and the acoustic driver are synchronized using trigger pulses provided by the imager. The phase offset (θ) between the two can be adjusted to acquire wave images at different phases of the cyclic motion

Table 3.1 Imaging parameters of a GRE–MRE sequence in liver MRE application

Parameters	Typical values
Magnetic field strength	1.5T
Patient body habitus	Supine/feet first
Orientation	Axial
Field of view (FOV)	30~42 cm (Patient-specific)
Fractional phase FOV	0.7–1.0 (Patient-specific)
In-plane resolution	256×64
Frequency direction	R/L
TR/TE	50/20
Slice thickness	10 mm
Number of slices	4
Flip angle/number of echoes/ETL	30/1/1
NEX/number of shots	1/0
ASSET acceleration factor	2
Active driver (resoundant system) frequency	60 Hz
Active driver (resoundant system) power	50 % power
MEG frequency/period	60 Hz/16.7 ms
Number of MEG gradient pairs	1
MEG amplitude	1.76 Gauss/cm (Machine dependent)
MEG direction	slice direction
Number of phase offsets	4
Acquisition time	54 s or less (four breath-hold)

angle$=30°$, slice thickness$=10$ mm, number of slices$=4$, TR/TE$=50/20$ ms, parallel imaging factor$=2$, 4 evenly spaced phase offsets, and 1 pair of 60 Hz trapezoidal MEGs with zero[th] and first moment nulling along the through-plane direction. Two spatial presaturation bands are applied on each side of the selected slice to reduce motion artifacts from blood flow. The total acquisition time is less than a minute, split into 4 breath holds.

Several artifact reduction strategies are used to maintain high image quality. In order to obtain a consistent position of the liver and establish more reproducible breath-hold positions for each suspended respiration, subjects are asked to hold their breath at the end of expiration. Breath-hold technique scan time is shorter than would likely be achieved with a respiratory gated sequence [39]. In the application of MRE with passive drivers, bulk motion can induce signal loss, ghosting, and unwanted phase accumulation, as shown in the first row of Fig. 3.3. However, we want spins to accrue phase from applied sinusoidal motion only. In order to separate these two sources of phase accumulation, Gradient Moment Nulling (GMN) can be applied to the motion sensitizing gradient waveform. It is well known that so called 1-2-1 (pulse widths of the first and last lobe are half of the central lobe) gradient pulses inherently have their 0th and 1st moments nulled. By definition, moments are a linear operation. Therefore, multiple 1-2-1 pulses can be superimposed on each other to yield longer trains of GMN pulses. As shown in Fig. 3.2, one pair of bipolar motion sensitizing gradients with GMN had their 0th and 1st moments nulled without changing the sensitivity to the desired sinusoidal shear motions. Additionally, nulling the 1st moment of the motion sensitizing gradient waveform does not require the entire sequence to have its 1st moment nulled. Therefore, as shown in the second row of Fig. 3.3, this reshaping of the motion sensitizing gradient waveforms to null the 1st moment reduces their susceptibly to bulk motion without increasing the echo time or reducing sensitivity to cyclic motion. Spatial presaturation bands were placed in above and below the liver to saturate possible incoming blood signal from descending aorta and inferior vena cava as shown in the third row of Fig. 3.3.

Intra-Voxel Phase Dispersion

The MEG gradients generate a phase accumulation within an isochromat. In areas of large displacements near the motion source, the phase across an isochromat can begin to disperse. This process is illustrated in Fig. 3.4. The signal dropout can be partially ameliorated by increasing the sampling resolution at the cost of increased breath-hold time. Alternatively, the driver amplitude can be reduced. However this decreases wave penetration and the phase-to-noise ratio. Some signal dropout is an acceptable trade-off to

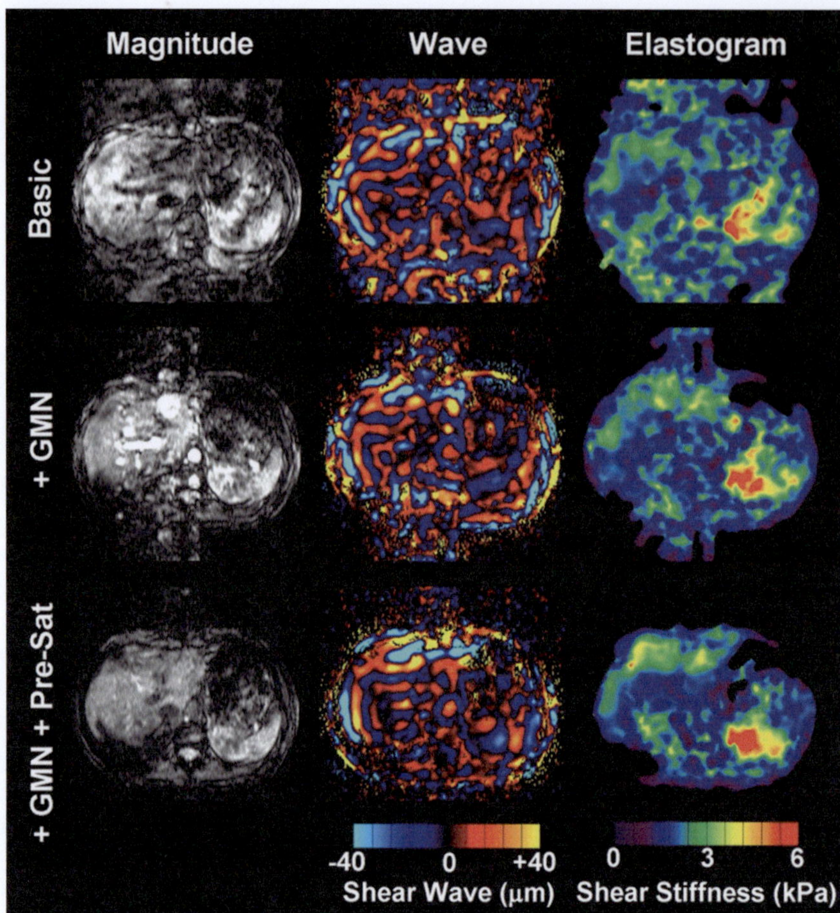

Fig. 3.3 Gradient echo MR elastography with gradient moment nulling and presaturation. This experiment demonstrated that bulk motion can contaminate both the magnitude and wave images and lead to unreliable shear stiffness, as shown in the *upper row*. The *middle row* illustrates results implemented with gradient moment nulling. Obviously, the bulk motion was mostly removed from the wave images. To further clean up the motion artifacts from blood flow, presaturation was implemented as well to improve the MRE wave image quality as shown in the *bottom row*

provide sufficient wave penetration into the liver. It is important to exclude these areas of signal void from analysis of liver stiffness.

Alternate Pulse Sequences

Other MRE sequences are available for evaluation of liver stiffness. EPI based Elastography sequences offer significant increases in speed of acquisition over GRE sequences. In multi-slice applications a spin echo implementation, shown in Fig. 3.5, can efficiently sample data for an Elastography study. Patients who have high iron concentration in their liver tissue, which leads to severely, shortened T2*, can be problematic at 1.5 and 3.0 T because it produces poor MR signal in the liver with GRE MRE sequence. Valid stiffness measurements may not be obtained due to the extremely low SNR. Since the signal in SE-EPI sequences experience T2 decay instead of the faster T2* decay, SE-EPI sequences may be an option for patients with high concentrations of iron.

To make the signal further less sensitive to the reduced T2*, which can be as low as ~1.3 ms in iron overloaded liver, a short-TE spin echo and spin echo EPI pulse sequence have been

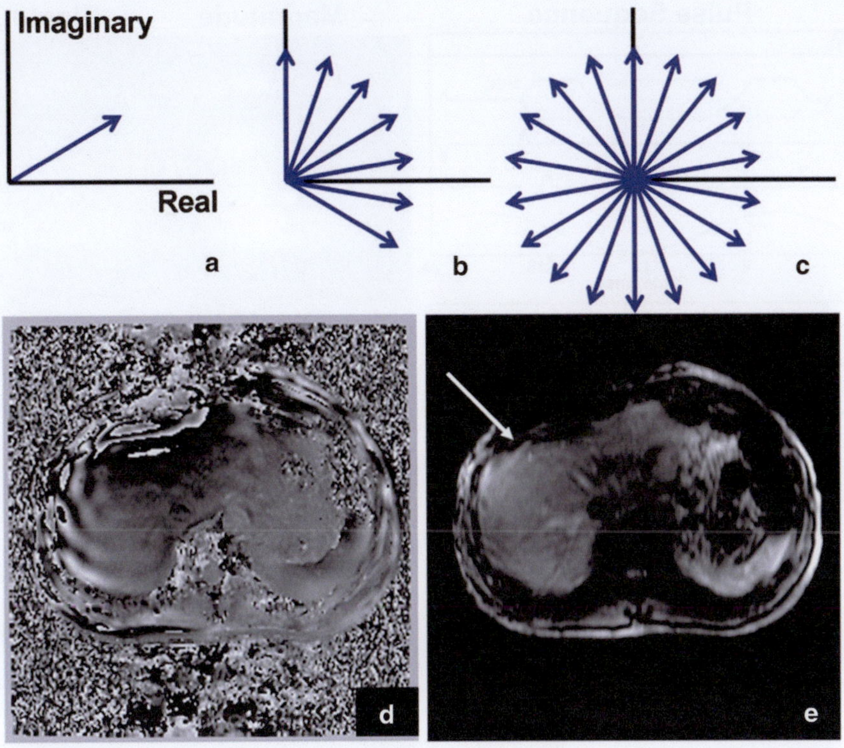

Fig. 3.4 Magnetization vectors (*blue arrows*) in the rotating reference frame for a spin isochromat on resonance during application of MRE pulse. (**a**) Depicts the signal in the complex plane of spins within a voxel not experiencing the effects of motion. As motion increases, the spins begin to dephase (**b**). Complete signal loss can occur in areas of high motion and the resulting high phase accumulation (**c–e**) (color figure online)

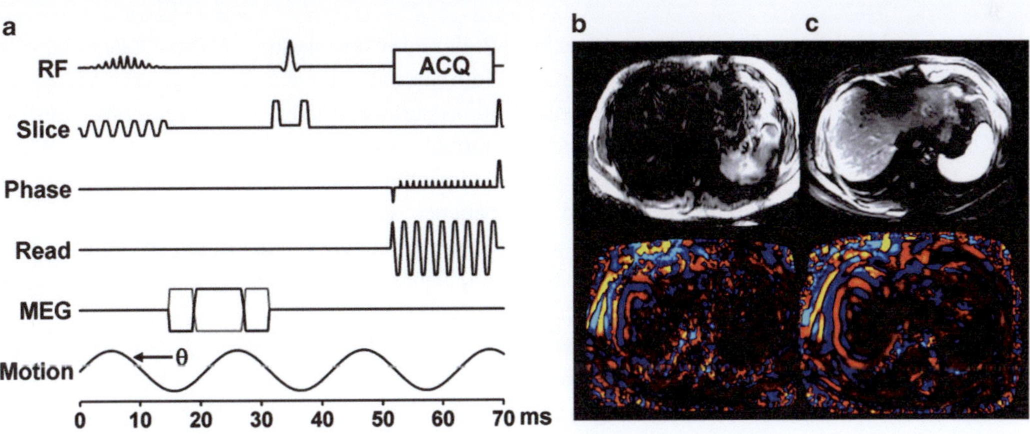

Fig. 3.5 Spin Echo EPI (SE-EPI) pulse sequence. Due to chemical shift artifacts, spatial-spectral pulses are used to generate the initial 90° RF pulse (**a**). Magnitude and wave images obtained at 3T from a GRE sequence (**b**) and the SE-EPI sequence (**c**) show increased signal and better wave depiction obtained with the EPI sequence. Susceptibility artifacts associated with EPI based sequences are not a problem in the liver application

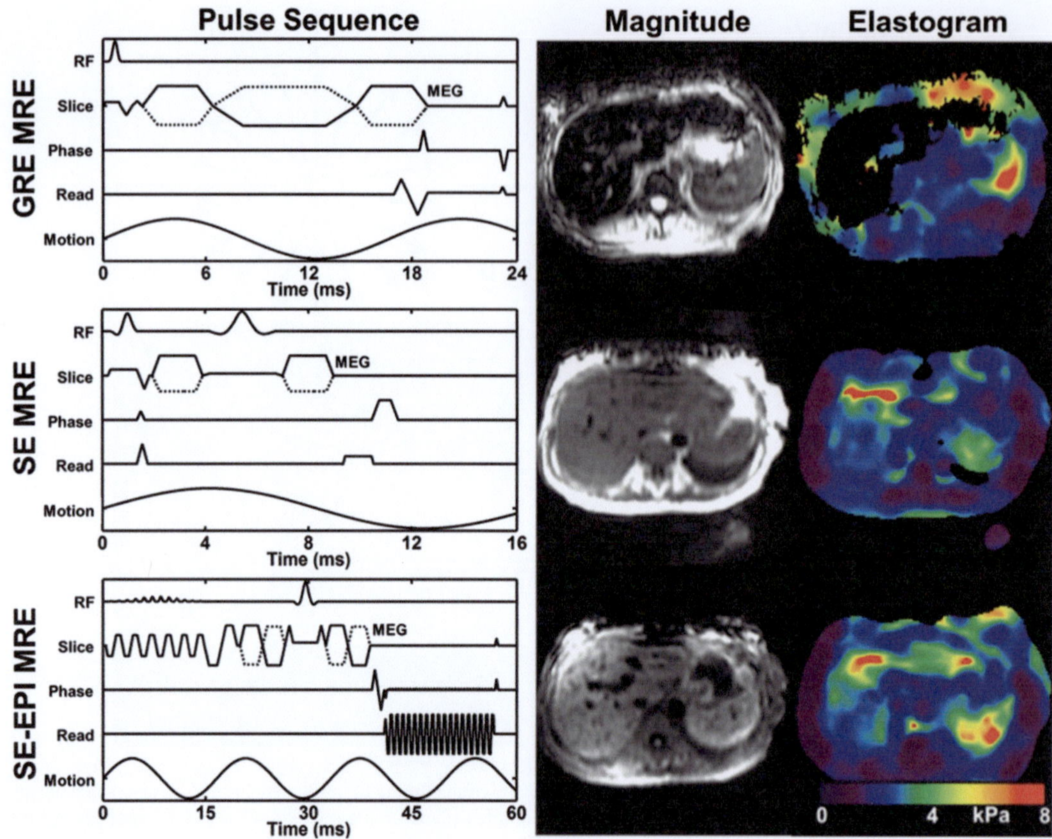

Fig. 3.6 The three pulse sequences and the corresponding MRE data in a patient with iron overload. The GRE technique (*top row*) failed, but the SE (*middle row*) and EPI (*bottom row*) acquisitions provided valid elastograms

developed and evaluated [56, 57]. The spin echo sequence modification includes fractional MEGs, crusher gradient removal, and split-unipolar MEGs of 2 ms duration positioned on either side of the 180° RF pulse. With these developments, the TE of the sequence could be reduced to 10 ms, and a schematic representation of this sequence is shown in the middle row of Fig. 3.6. In addition to the SE MRE sequence, a spin echo EPI pulse sequence was also developed allowing for faster imaging with modifications similar to the SE MRE pulse sequence. The advantages and disadvantages of EPI-specific parameters including the number of shots and the effective TE were optimized for liver MRE specifically. To achieve the low TE value, chemical presaturation pulses were used instead of a spatial-spectral RF pulse for fat suppression. A schematic of this pulse sequence is shown in the bottom row of Fig. 3.6.

Preliminary evaluation of these techniques on both normal volunteers and patients with known iron overloaded liver is very promising [36]. In normal livers, all three sequences provided comparable shear stiffness measurements. In patients, the significant increase of the liver parenchyma signals and high confidence of the calculated stiffness values can be appreciated from SE and SE-EPI MRE sequences.

STEAM based sequences, seen in Fig. 3.7, are also an option and have been successfully used in cardiac imaging [58, 59]. Most implementations use non-selective hard pulses for the first two RF pulses. This requires a long TR for the signal to recover. However, STEAM Elastography pulse sequences can be efficient in TE. The MEG gradients do not need to be turned on over and entire motion cycle. They can be split and turned on only during a small fraction of the motion cycle.

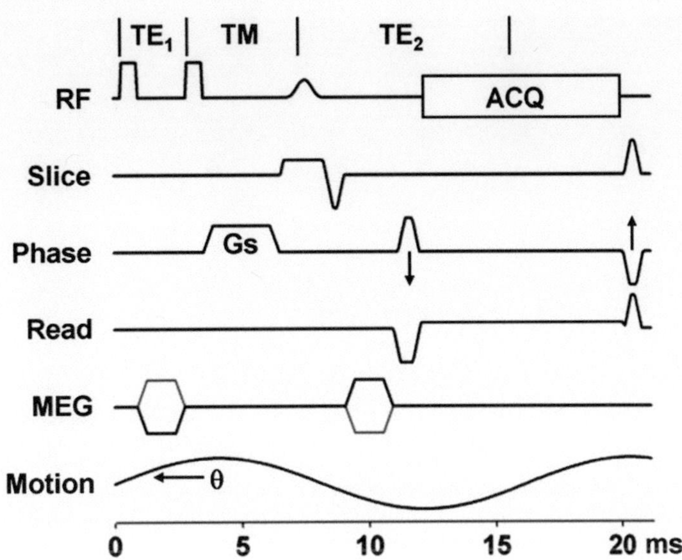

Fig. 3.7 STEAM sequences inherently split the signal pathways so that only half of the signal is available to sample. The advantage is that T2 decay only occurs during the TE$_1$ and TE$_2$ periods. During the TM period, spins not originating with the first RF can be spoiled with the Gs gradient. Spins following the STEAM pathway will be in the longitudinal plane and will not be affected by the Gs gradient or T2 decay. The MEG gradients are still synchronized to the applied motion however the spins spend much less time in the transverse plan experiencing T2 decay

Slice Selection Criteria for 2-D Liver MRE Acquisition

The MRE visualization of the full 3-D vector displacement field within the liver produced by the passive drivers has proven to offer insights and improvements in both experimental design and 2-D imaging plane selection for in vivo MRE applications [59, 60]. When the results of the 3-D liver MRE acquisition and inversions were used as "ground truth," 2-D imaging in the "widest" cross-section of the liver consistently yielded inversion results that were comparable to those obtained with the much more time consuming 3-D imaging approach [61]. These studies showed that in a large portion of the liver, the propagation direction of shear waves is approximately parallel with the transverse imaging plane when using a pressure-activated passive driver on the anterior chest wall. This allows accurate elastograms to be generated from 2-D wave image data (which can be acquired in one breath-hold), rather than 3-D wave imaging, which requires longer acquisition times and more complex measures to address respiratory motion. With the acoustic driver system, the 2-D wave approximation is generally valid in the axial imaging planes distanced 2–10 cm away from the superior margin of the liver, assuming the liver has average vertical height of 15 cm. It corresponds to a region across the widest part of the liver as shown in Fig. 3.8. In other marginal levels/regions where the 2-D planar wave model may frequently break down due to special boundary conditions, the shear waves usually traverse obliquely through the plane of interest and may result in erroneous stiffness values.

Software (Inversion Algorithms)

Dynamic MRE uses propagating mechanical shear waves rather than a static stress as a means to study tissue elasticity. This provides an important advantage for calculating elasticity in that dynamic MRE does not require the estimation of the regional static stress distribution either inside or on the boundaries of the tissue. The acquired wave images are processed using an inversion algorithm to generate quantitative images that depict tissue stiffness, called elastograms. Many different inversion algorithms have been developed to process MRE wave data based on different assumptions about the nature of the wave propagation. Pre-processing algorithms for the wave data include phase unwrapping, removal of concomitant gradient field effects, and directional filtering [62] to enhance the accuracy of the elastograms. Additional pre-processing generates

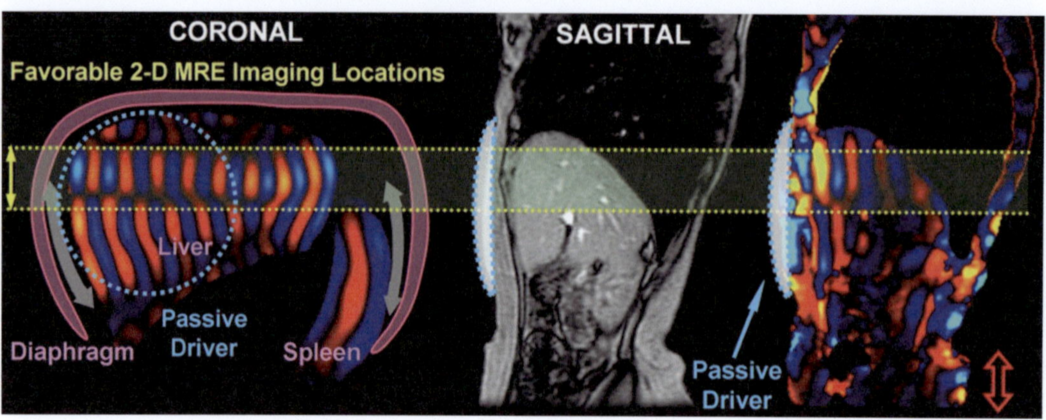

Fig. 3.8 Schematic illustrations of the hepatic MRE setup (with passive driver position indicated) and shear wave generation mechanism (diaphragm act as a secondary driver). Favorable 2-D Liver MRE slice selection locations are indicated in the coronal simulation shown on the *right*. This was validated in a human study with sagittal imaging locations (motion sensitizing direction is S/I)

a Background Phase Estimate (BPE) that includes the longitudinal waves, which can be removed to provide the shear wave signal. The inversion algorithms produce stiffness maps using one of the following: measures of the phase gradient or spatial frequency content, direct inversions of the differential equations of motion, or iterative methods involving finite element models [63–67].

The process of generating a stiffness map is achieved by using the MRE displacement data in sliding windows to perform a direct inversion (DI) of the differential equation modeling the wave propagation. A second-order or fourth-order polynomial is fit to the data depending on fit quality. The correlation coefficient (R^2) for the fit is recorded for the center pixel of each window. This results in a confidence map with values between 0 and 1, the latter representing high confidence and good signal-to-noise ratio (SNR) [67, 68]. The phase difference wave data obtained from MRE may include the effects of both shear and longitudinal waves. Prior to processing with DI inversion algorithm, an 8-direction motion filters [62] evenly spaced between 0° and 360° and combined in a weighted least-square method to improve the performance of the algorithm are applied, since complex interference of shear waves from all directions may produce areas with low shear displacement amplitude. A BPE is removed from each of the eight directionally filtered data sets. A DI algorithm is then used to estimate the stiffness using the shear wave estimate obtained with each region. SNR map is also obtained using both the magnitude and phase images obtained during the scan. Assuming a constant signal, the magnitude SNR is calculated over a small processing kernel from all the phase offsets. As shown in Fig. 3.9, the phase SNR is obtained

with: $SNR_{phase} = | \text{shearwave} | / \sqrt{2} \sin^{-1} \left(SNR_{mag} \right)$.

Reliability of the liver stiffness measurement at each voxel is then determined by the confidence values from SNR maps.

All of these steps in processing are done automatically, without human intervention, to yield quantitative images of tissue shear stiffness, in units of kilopascals (kPa). We have used the designation "shear stiffness," rather than "shear modulus" to indicate that the measurements are effective stiffness values at the driving frequency, and may include a viscous component, although this is likely to be small at the low driving frequency used.

The stiffness maps (elastograms) thus generated include the entire cross-section of the abdomen including the liver and other abdominal organs. Stiffness values can be obtained by placing regions of interest (ROI) on any organ included in that section.

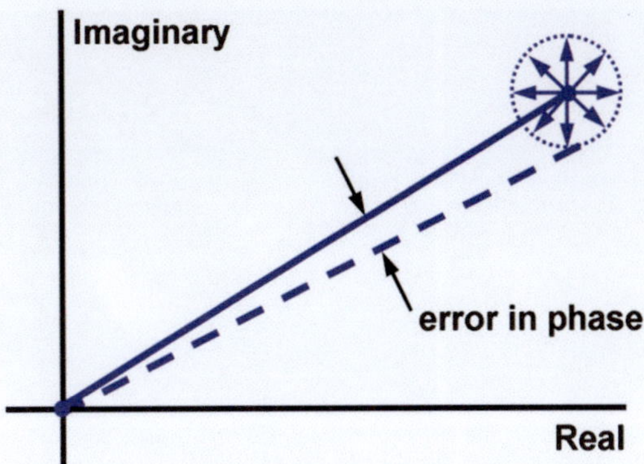

Fig. 3.9 Assuming that the MR signal has normally distributed noise in the complex plane, the phase SNR is obtained as a function of arc sin (1/magnitude SNR) and shear wave amplitude. The value is divided by 2, since we gain two factors of $\sqrt{2}$. One $\sqrt{2}$ is due to the noise vector being at an arbitrary angle to the signal vector and another $\sqrt{2}$ is from sampling two phase images with opposite gradient polarity and taking their difference (thus adding their signals)

Analysis of Stiffness Maps (ROI Placement for Stiffness Measurement)

The elastograms are analyzed by measuring mean shear stiffness within a large, manually specified region of interest that includes hepatic parenchyma, while excluding major blood vessels, such as hepatic veins, main portal veins, and branches that have width greater than 6 pixels (about 8 mm); liver fissures, gall bladder fossa, and edge of the liver. 2-D liver elastograms are usually calculated based on the assumption of a 2-D planar shear wave field described previously. Ideally, the entire liver parenchyma should be evenly illuminated by planar shear waves with adequate amplitude. However, in practice, the shear waves can be quickly attenuated in the deep structure of the liver; or appear separated in different regions of the liver due to destructive/constructive wave interference at intersections. Therefore, not all hepatic tissue in the final liver stiffness map is measurable. Regions without adequate magnitude signal or shear wave amplitude, which will be identified with the confidence map (as shown in Fig. 3.10), can result in erroneous stiffness value. It is crucial to obtain accurate mean liver stiffness values with appropriate placement of the ROI. Analysis of the liver elastogram should be performed with all original magnitude, phase difference (wave) images, and automatically calculated confidence maps (Fig. 3.10) to ensure that the calculated liver stiffness is reliable.

Due to the complexity of selecting an artifact-free ROI in MRE and the subjectivity of evaluating the wave interference, the hepatic stiffness values calculated by multiple analysts from the same set of images can differ. Tools which reduce inter- and intra-reader variability (e.g., the confidence map) are vital to the standardization of MRE measurements. Their development is made complicated, however, by the motion artifact in the images, the respiratory mismatch across slices and phase offsets, as well as the low contrast, which are all common in MRE.

An algorithm which calculates stiffness in a fully automated manner is presented in [68]. It has been cross validated to perform as reliably and accurately as well-trained MRE image analysts under typical imaging conditions and is undergoing prospective evaluation. Unlike manual analysis, which often results in smaller ROIs representative of the liver being selected, the automated tool aims to include all reliable parenchyma (Fig. 3.11) which leads to better sampling and provides the potential for more sophisticated analysis than global stiffness averaging. For these reasons, it can offer significant advantages over manual techniques for reading MRE cases in terms of efficiency, repeatability, and reproducibility.

Customized image analysis software (MRE_Quant software) for dedicated analysis of the liver MRE and reporting is also available. The ROI drawing steps and criteria are listed as follows and illustrated in Fig. 3.12 as well.

(1) Open the software, load all MRE data, including both magnitude and wave images. The corresponding elastograms and confidence maps (i.e., SNR maps) will be automatically calculated before analysis.

(2) If it is a longitudinal study, load other images acquired at different time points by opening a series of separate windows. Use the MRE magnitude image at the first time point as the reference and register other time point images to it.

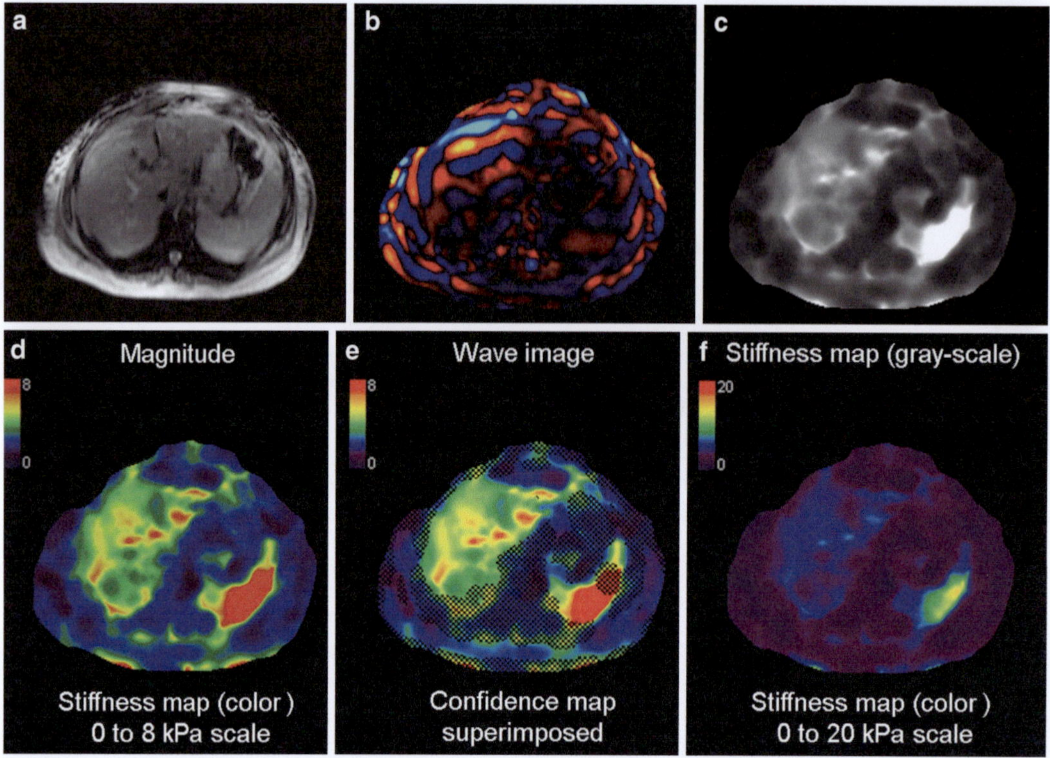

Fig. 3.10 Typical set of images obtained with 2D GRE MRE sequence for a single slice. Magnitude image (**a**), color phase wave image (**b**), the gray scale stiffness map (**c**) on which measurements are obtained, color stiffness map with 0–8 kPa color scale (**d**), stiffness map with confidence map superimposed (**e**), and color stiffness map with 0–20 kPa color scale (**f**). The stiffness values are obtained on the gray scale stiffness map. Typically on multi-monitor or multi-layout workstations ROIs can be first drawn on any of these images and copy pasted onto others. Other set of images including gray scale phase image, stiffness map with confidence map superimposed on 0–20 kpa color scale are not illustrated here

(3) On the MRE images, draw and modify a closed, connected ROI contour in the liver region, about several pixels away from the liver boundary, excluding vessels and non-hepatic tissues, and avoiding regions of severe wave interferences. The software will automatically accept only those points within the ROI that also have confidence values above a specific threshold (e.g., 0.95 as a default threshold). Points with confidence values below this threshold will be masked with a checkerboard pattern in the elastogram.

(4) Repeat step (3) for other slices or other time point data in a similar way.

(5) Save every individual ROIs and/or intersection ROIs (for longitudinal study). Based on well-defined and saved ROIs, enable the statistical analysis function to report liver stiffness measurements and related statistics (mean, standard deviation, mode, histogram, etc.) in an output Excel file. A Microsoft Office PowerPoint (PPT) file, containing all MRE images/animations and liver stiffness measurements, can also be created to facilitate a quick reference in the future.

Placing ROIs and Avoiding Errors

Some important considerations are needed when placing ROIs for the measurement of liver stiffness in clinical application of 2-D Liver MRE. First of all, non-hepatic tissue, such as fissures, gall bladder fossa, vessels etc., should be excluded as show in Fig. 3.13.

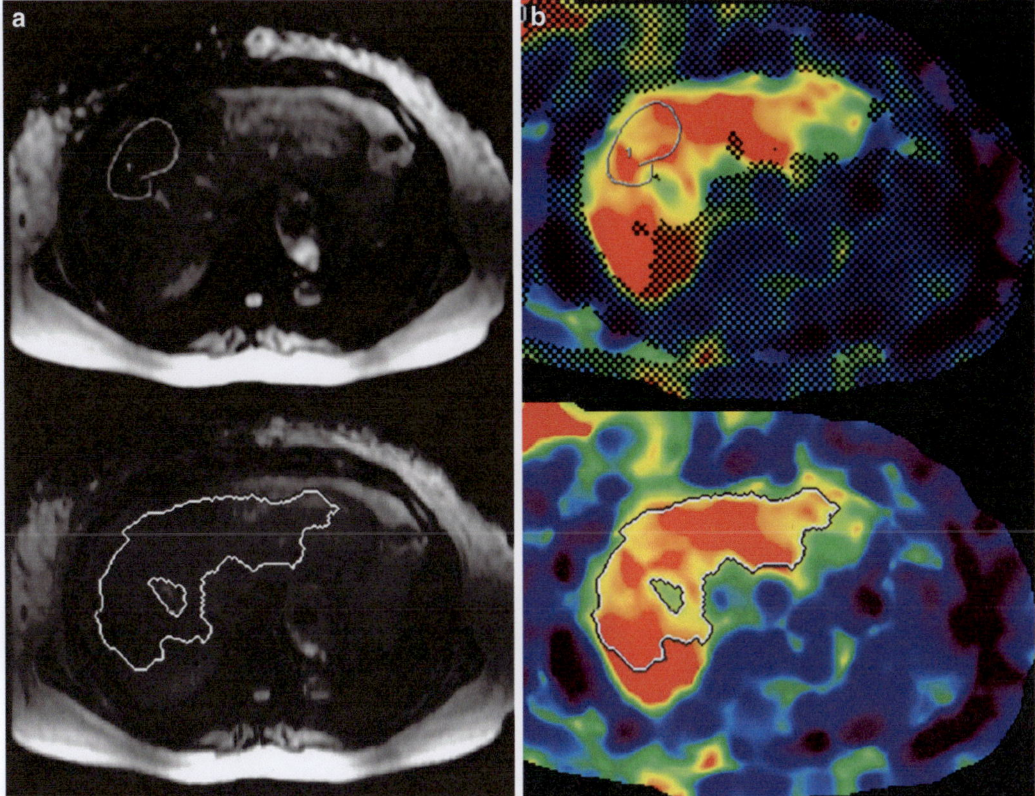

Fig. 3.11 The interfaces and outputs of the Liver MRE analysis software MRE/ROI. (**a**) The left interface image illustrates the procedure of placing and editing a ROI on a single hepatic MRE data set. The deep structure of hepatic tissues was excluded due to insufficient wave amplitude (nearly invisible from appearance). The *right* interface image shows a series of longitudinal studies (four time points) with magnitude images (*top row*), wave images (*central row*), and confidence maps (*bottom row*). Reliable regions include hepatic tissues illuminated with visible planar waves and high confidence values. In order to minimize the subjectivity, investigators are blinded to the elastograms during ROI drawing. (**b**) An output Microsoft Office PowerPoint file to demonstrate hepatic MRE images, defined ROIs and liver stiffness measurements. The ROI selected in the top row case illustrates the criteria of excluding non-hepatic tissue, while the *bottom* ROI shows the criteria of excluding an intersection region with destructive wave interference

Ideally, the entire liver parenchyma should be analyzed with MRE technique. However, extra care must be taken to address the limitations of 2-D MRE wave imaging and processing method. Nonplanar shear waves, such as wave interference (cancellation "−" or prolongation "+") and obliquity, should be excluded as shown in Fig. 3.14.

Preparing for a Liver MRE Study

Food intake is known to cause an increase in mesenteric blood flow, which may lead to a postprandial increase in hepatic stiffness that is different in patients with hepatic fibrosis than in normal volunteers [69]. It has been observed that MRE-assessed liver stiffness increases significantly (ranges from 5 up to 48 %) following a test meal in patients with advanced hepatic fibrosis, whereas no significant changes occur in a healthy normal individuals [69]. This finding suggests that there is a dynamic component to the liver stiffness that is dependent on the increased perfusion. Therefore, it is very important to have hepatic MRE examinations performed consistently in a fasting state. The postprandial augmentation in hepatic stiffness after a test meal known to increase mesenteric

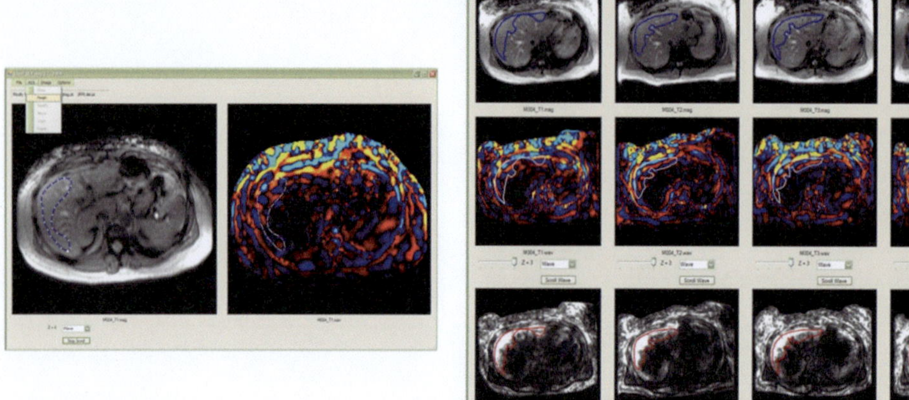

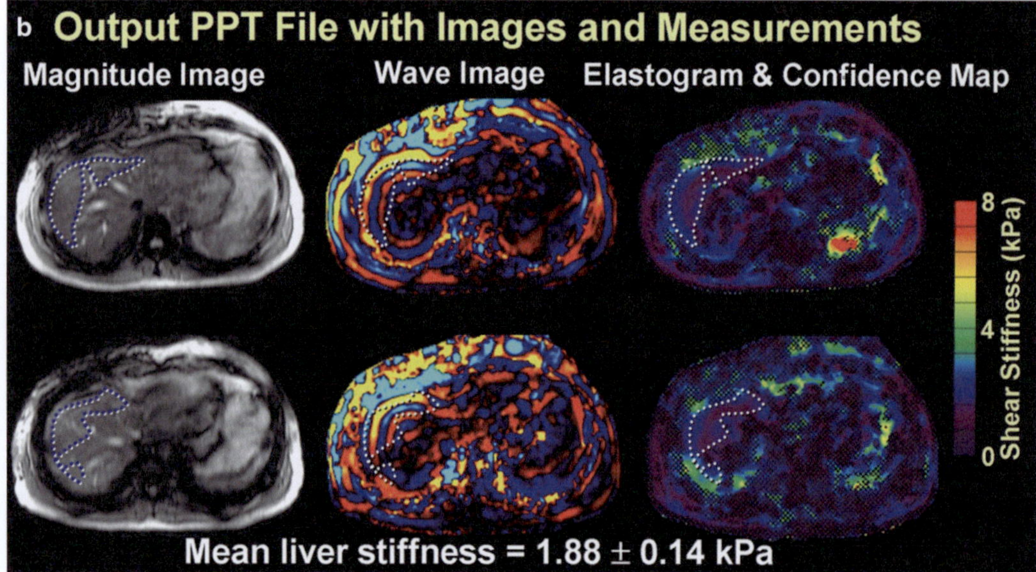

Fig. 3.12 Representative ROIs generated by an experienced clinical MRE analyst (*top row*) and by the automated method (*bottom row*) are shown. Both approaches use the magnitude, wave, and stiffness information to select an artifact-free region of liver parenchyma. The automated method typically provides a more thorough sampling by including all reliable tissue. The checkerboard represents areas excluded from the manual ROI due to low values of the confidence map which is related to SNR

blood flow is likely due to the transiently increased portal pressure in patients with hepatic fibrosis. It is thought that mechanical distortion of the intrahepatic vasculature caused by fibrosis impairs the autoregulatory mechanism for the portal venous pressure, which may cause acceleration of the development of portal systemic varies and stretching of hepatic parenchyma and stellate cells that are instrumental in the progression of hepatic fibrosis. This promising observation

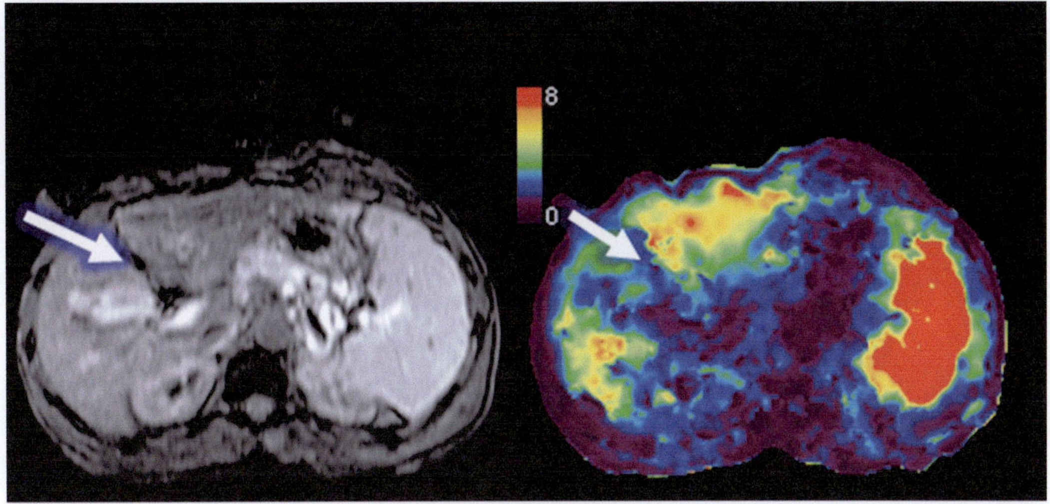

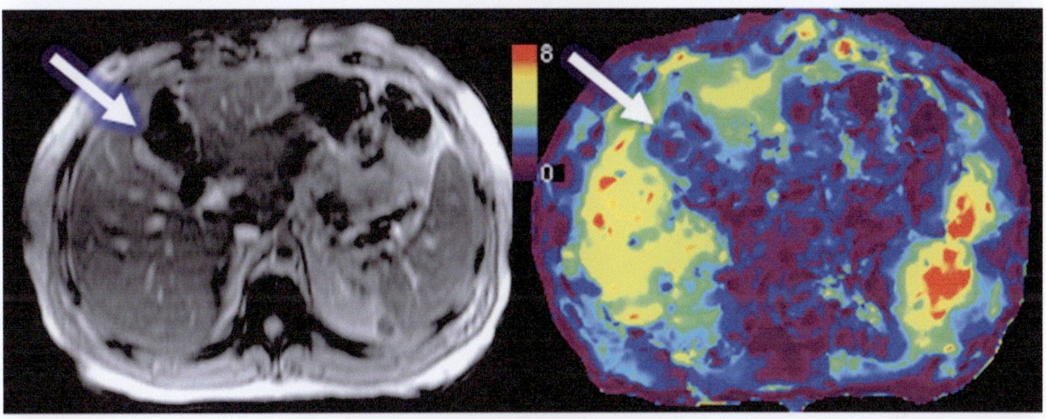

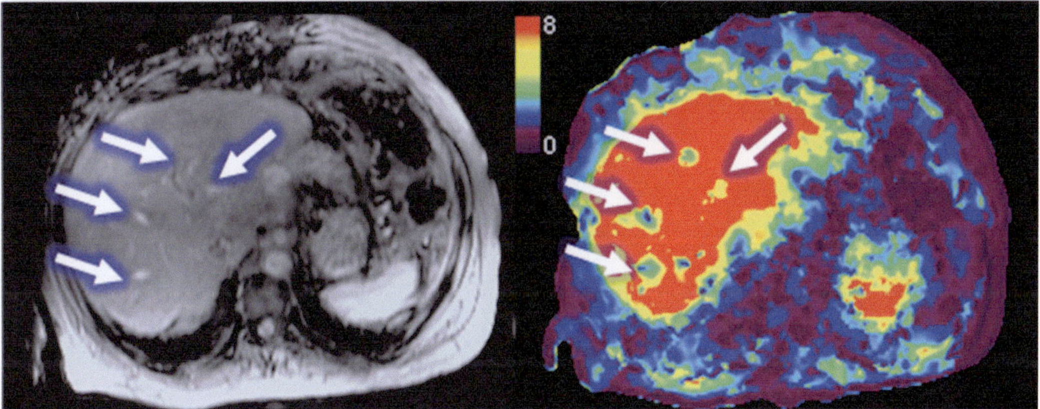

Fig. 3.13 Illustration of a criterion of exclusion of non-hepatic tissue for liver stiffness measurements. A case in *top row*: low stiffness areas corresponding to fissures; A case in *middle row*: a low stiffness area corresponding to gall bladder fossa; A case in *bottom row*: low stiffness areas corresponding to vessels in cirrhotic liver

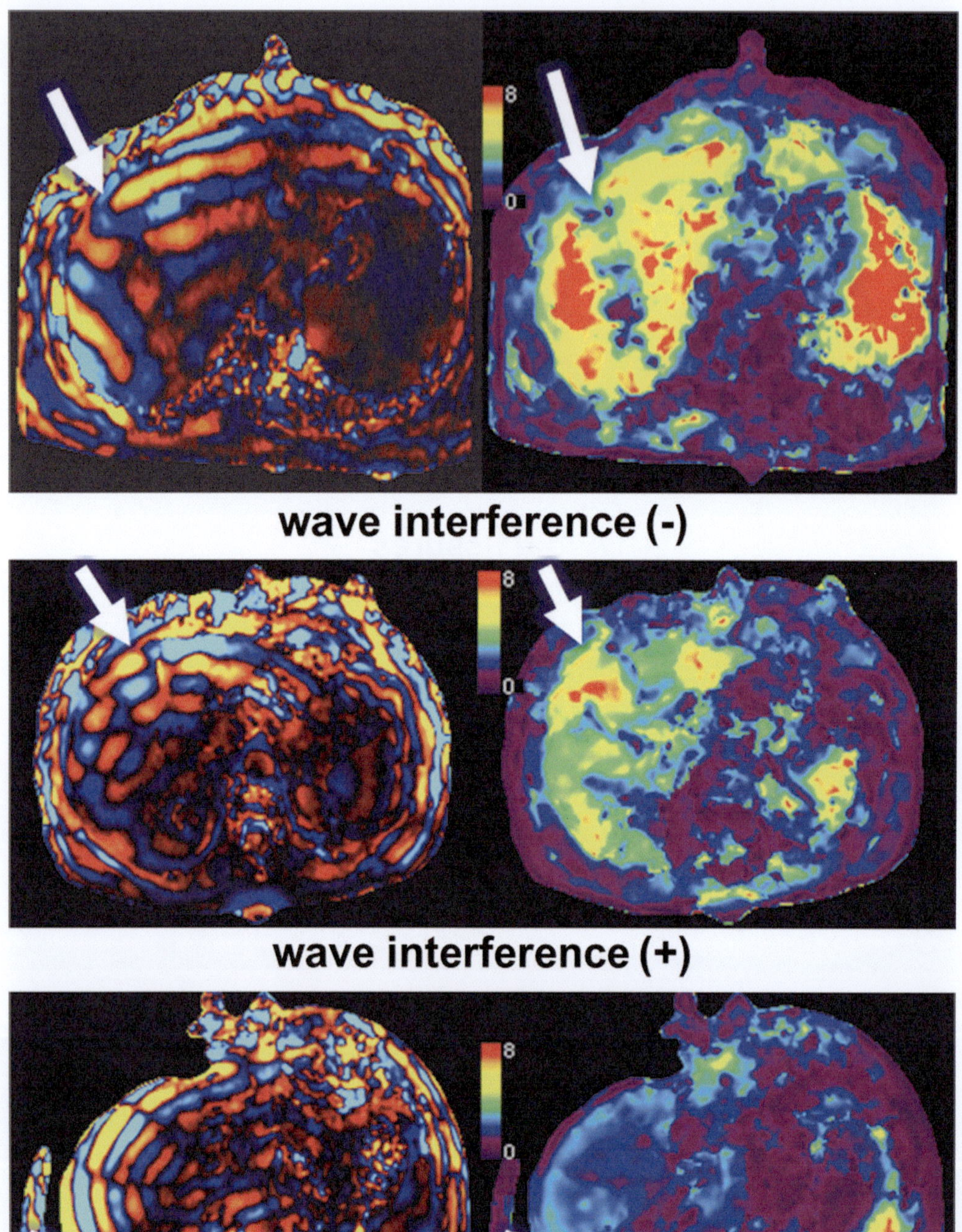

Fig. 3.14 Illustration of the criteria of exclusion of wave interference and obliquity for liver stiffness measurements in 2-D wave data. A case in *top row*: wave interference resulting in subtraction "−" effect leading to low stiffness areas; A case in *middle row*: wave interference resulting in addition "+" effect leading to higher stiffness areas ; A case in *bottom row*: Normal liver showing wave obliquity effect in posterior part of the liver resulting in higher stiffness values in elastogram

provides motivation for further studies to determine the potential value of assessing postprandial hepatic stiffness augmentation for predicting progression of fibrotic disease and the development of portal varices. It may also provide new insights into the natural history and pathophysiology of chronic liver disease.

Conclusion

Numerous studies have shown that it is readily possible to perform MRE of Liver. Motivated by the results of the pilot volunteer and patient studies [39, 43, 44], multiple investigators have studied the diagnostic ability of liver MRE for detecting hepatic fibrosis in patients with chronic liver disease from various etiologies [43, 45, 70–72]. The results from these studies have established that liver MRE has excellent diagnostic accuracy for assessing hepatic fibrosis, which will be discussed in Chap. 4. Motivated by the successful implementation of MRE for the study of diffuse changes in hepatic stiffness due to fibrosis, studies has been conducted to evaluate the potential role of MRE in characterizing hepatic tumors, which will be discussed in Chap. 5.

References

1. Schuppan D, Popov Y. Hepatic fibrosis: from bench to bedside. J Gastroenterol Hepatol. 2002;17 Suppl 3:S300–5.
2. Tsukada S, Parsons CJ, Rippe RA. Mechanisms of liver fibrosis. Clin Chim Acta. 2006;364(1–2):33–60.
3. Fowell AJ, Iredale JP. Emerging therapies for liver fibrosis. Dig Dis. 2006;24(1–2):174–83.
4. Friedman SL, Bansal MB. Reversal of hepatic fibrosis – fact or fantasy? Hepatology. 2006;43(2 Suppl 1):S82–8.
5. Talwalkar JA, Yin M, Fidler JL, Sanderson SO, Kamath PS, Ehman RL. Magnetic resonance imaging of hepatic fibrosis: emerging clinical applications. Hepatology. 2008;47(1):332–42.
6. Salvatore V, Borghi A, Peri E, Colecchia A, Li Bassi S, Montrone L, et al. Relationship between hepatic haemodynamics assessed by Doppler ultrasound and liver stiffness. Dig Liver Dis. 2012;44(2):154–9 [Comparative Study].
7. Lerner RM, Huang SR, Parker KJ. "Sonoelasticity" images derived from ultrasound signals in mechanically vibrated tissues. Ultrasound Med Biol. 1990; 16(3):231–9.
8. Sandrin L, Tanter M, Gennisson JL, Catheline S, Fink M. Shear elasticity probe for soft tissues with 1-D transient elastography. IEEE Trans Ultrason Ferroelectr Freq Control. 2002;49(4):436–46.
9. Muthupillai R, Lomas DJ, Rossman PJ, Greenleaf JF, Manduca A, Ehman RL. Magnetic resonance elastography by direct visualization of propagating acoustic strain waves. Science. 1995;269(5232):1854–7.
10. Bohte AE, de Niet A, Jansen L, Bipat S, Nederveen AJ, Verheij J, et al. Non-invasive evaluation of liver fibrosis: a comparison of ultrasound-based transient elastography and MR elastography in patients with viral hepatitis B and C. Eur Radiol. 2013;25.
11. Bensamoun SF, Wang L, Robert L, Charleux F, Latrive JP, Ho Ba Tho MC. Measurement of liver stiffness with two imaging techniques: magnetic resonance elastography and ultrasound elastometry. J Magn Reson Imaging. 2008;28(5):1287–92.
12. Huwart L, Sempoux C, Vicaut E, Salameh N, Annet L, Danse E, et al. Magnetic resonance elastography for the noninvasive staging of liver fibrosis. Gastroenterology. 2008;135(1):32–40 [Clinical Trial Comparative Study Research Support, Non-U.S. Gov't Validation Studies].
13. Oudry J, Chen J, Glaser KJ, Miette V, Sandrin L, Ehman RL. Cross-validation of magnetic resonance elastography and ultrasound-based transient elastography: a preliminary phantom study. J Magn Reson Imaging. 2009;30(5):1145–50.
14. Sandrin L, Fourquet B, Hasquenoph JM, Yon S, Fournier C, Mal F, et al. Transient elastography: a new noninvasive method for assessment of hepatic fibrosis. Ultrasound Med Biol. 2003;29(12):1705–13.
15. Fraquelli M, Rigamonti C, Casazza G, Conte D, Donato MF, Ronchi G, et al. Reproducibility of transient elastography in the evaluation of liver fibrosis in patients with chronic liver disease. Gut. 2007; 56(7):968–73.
16. Castera L, Vergniol J, Foucher J, Le Bail B, Chanteloup E, Haaser M, et al. Prospective comparison of transient elastography, fibrotest, APRI, and liver biopsy for the assessment of fibrosis in chronic hepatitis C. Gastroenterology. 2005;128(2):343–50.
17. Ziol M, Handra-Luca A, Kettaneh A, Christidis C, Mal F, Kazemi F, et al. Noninvasive assessment of liver fibrosis by measurement of stiffness in patients with chronic hepatitis C. Hepatology. 2005;41(1): 48–54.
18. Foucher J, Chanteloup E, Vergniol J, Castera L, Le Bail B, Adhoute X, et al. Diagnosis of cirrhosis by transient elastography (FibroScan): a prospective study. Gut. 2006;55(3):403–8.
19. Ganne-Carrie N, Ziol M, de Ledinghen V, Douvin C, Marcellin P, Castera L, et al. Accuracy of liver stiffness measurement for the diagnosis of cirrhosis in patients with chronic liver diseases. Hepatology. 2006;44(6):1511–7.

20. Coco B, Oliveri F, Maina AM, Ciccorossi P, Sacco R, Colombatto P, et al. Transient elastography: a new surrogate marker of liver fibrosis influenced by major changes of transaminases. J Viral Hepat. 2007;14(5): 360–9.

21. Kruse SA, Rose GH, Glaser KJ, Manduca A, Felmlee JP, Jack Jr CR, et al. Magnetic resonance elastography of the brain. Neuroimage. 2008;39(1):231–7.

22. Green MA, Bilston LE, Sinkus R. In vivo brain viscoelastic properties measured by magnetic resonance elastography. NMR Biomed. 2008;21(7):755–64.

23. Sack I, Beierbach B, Hamhaber U, Klatt D, Braun J. Non-invasive measurement of brain viscoelasticity using magnetic resonance elastography. NMR Biomed. 2008;21(3):265–71.

24. McKnight AL, Kugel JL, Rossman PJ, Manduca A, Hartmann LC, Ehman RL. MR elastography of breast cancer: preliminary results. AJR Am J Roentgenol. 2002;178(6):1411–7.

25. Van Houten EE, Doyley MM, Kennedy FE, Weaver JB, Paulsen KD. Initial in vivo experience with steady-state subzone-based MR elastography of the human breast. J Magn Reson Imaging. 2003;17(1): 72–85.

26. Sinkus R, Siegmann K, Xydeas T, Tanter M, Claussen C, Fink M. MR elastography of breast lesions: understanding the solid/liquid duality can improve the specificity of contrast-enhanced MR mammography. Magn Reson Med. 2007;58(6):1135–44.

27. Plewes DB, Bishop J, Samani A, Sciarretta J. Visualization and quantification of breast cancer biomechanical properties with magnetic resonance elastography. Phys Med Biol. 2000;45(6):1591–610.

28. Kolipaka A, McGee KP, Araoz PA, Glaser KJ, Manduca A, Romano AJ, et al. MR elastography as a method for the assessment of myocardial stiffness: comparison with an established pressure–volume model in a left ventricular model of the heart. Magn Reson Med. 2009;62(1):135–40.

29. Sack I, Rump J, Elgeti T, Samani A, Braun J. MR elastography of the human heart: noninvasive assessment of myocardial elasticity changes by shear wave amplitude variations. Magn Reson Med. 2009;61(3): 668–77.

30. McGee KP, Hubmayr RD, Ehman RL. MR elastography of the lung with hyperpolarized 3He. Magn Reson Med. 2008;59(1):14–8.

31. Ringleb SI, Chen Q, Lake DS, Manduca A, Ehman RL, An KN. Quantitative shear wave magnetic resonance elastography: comparison to a dynamic shear material test. Magn Reson Med. 2005;53(5):1197–201.

32. Talwalkar JA, Yin M, Venkatesh S, Rossman PJ, Grimm RC, Manduca A, et al. Feasibility of in vivo MR elastographic splenic stiffness measurements in the assessment of portal hypertension. AJR Am J Roentgenol. 2009;193(1):122–7.

33. Shah NS, Kruse SA, Lager DJ, Farell-Baril G, Lieske JC, King BF, et al. Evaluation of renal parenchymal disease in a rat model with magnetic resonance elastography. Magn Reson Med. 2004;52(1):56–64.

34. Shi Y, Glser KJ, Venkatesh SK, Ben-Abraham EI, Ehman RL. Feasibility of using 3D MR elastography to determine pancreatic stiffness in healthy volunteers. J Magn Reson Imaging. 2014. Accepted.

35. Yin M, Chen J, Glaser KJ, Talwalkar JA, Ehman RL. Abdominal magnetic resonance elastography. Top Magn Reson Imaging. 2009;20(2):79–87.

36. Mariappan YK, Venkatesh SK, Glaser JK, McGee KP, Ehman RL, editors. MR Elastography of liver with iron overload: development, evaluation and preliminary clinical experience with improved spin echo and spin echo EPI sequences. The 21st annual conference of international society for magnetic resonance in medicine; 2013; Salt Lake City, Utah.

37. Venkatesh SK, Yin M, Ehman RL. Magnetic resonance elastography of liver: clinical applications. J Comput Assist Tomogr. 2013;37(6):887–96.

38. Venkatesh SK, Yin M, Ehman RL. Magnetic resonance elastography of liver: technique, analysis, and clinical applications. J Magn Reson Imaging. 2013;37(3):544–55.

39. Huwart L, Peeters F, Sinkus R, Annet L, Salameh N, ter Beek LC, et al. Liver fibrosis: non-invasive assessment with MR elastography. NMR Biomed. 2006; 19(2):173–9.

40. Salameh N, Peeters F, Sinkus R, Abarca-Quinones J, Annet L, Ter Beek LC, et al. Hepatic viscoelastic parameters measured with MR elastography: correlations with quantitative analysis of liver fibrosis in the rat. J Magn Reson Imaging. 2007;26(4):956–62.

41. Huwart L, Sempoux C, Vicaut E, Salameh N, Annet L, Danse E, et al. Magnetic resonance elastography for the noninvasive staging of liver fibrosis. Gastroenterology. 2008;135(1):32–40.

42. Asbach P, Klatt D, Hamhaber U, Braun J, Somasundaram R, Hamm B, et al. Assessment of liver viscoelasticity using multifrequency MR elastography. Magn Reson Med. 2008;60(2):373–9.

43. Klatt D, Asbach P, Rump J, Papazoglou S, Somasundaram R, Modrow J, et al. In vivo determination of hepatic stiffness using steady-state free precession magnetic resonance elastography. Invest Radiol. 2006;41(12):841–8.

44. Rouviere O, Yin M, Dresner MA, Rossman PJ, Burgart LJ, Fidler JL, et al. MR elastography of the liver: preliminary results. Radiology. 2006;240(2):440–8.

45. Yin M, Talwalkar JA, Glaser KJ, Manduca A, Grimm RC, Rossman PJ, et al. Assessment of hepatic fibrosis with magnetic resonance elastography. Clin Gastroenterol Hepatol. 2007;5(10):1207–13.

46. Venkatesh SK, Yin M, Glockner JF, Takahashi N, Araoz PA, Talwalkar JA, et al. MR elastography of liver tumors: preliminary results. AJR Am J Roentgenol. 2008;190(6):1534–40.

47. Yin M, Woollard J, Wang X, Torres VE, Harris PC, Ward CJ, et al. Quantitative assessment of hepatic fibrosis in an animal model with magnetic resonance elastography. Magn Reson Med. 2007;58(2):346–53.

48. Bernstein MA, Ikezaki Y. Comparison of phase-difference and complex-difference processing in

phase-contrast MR angiography. J Magn Reson Imaging. 1991;1:725–9.

49. Muthupillai R, Rossman PJ, Lomas DJ, Greenleaf JF, Riederer SJ, Ehman RL. Magnetic resonance imaging of transverse acoustic strain waves. Magn Reson Med. 1996;36(2):266–74.

50. Rump J, Klatt D, Braun J, Warmuth C, Sack I. Fractional encoding of harmonic motions in MR elastography. Magn Reson Med. 2007;57(2):388–95.

51. Glaser KJ, Felmlee JP, Ehman RL. Rapid MR elastography using selective excitations. Magn Reson Med. 2006;55(6):1381–9.

52. Sinkus R, Tanter M, Catheline S, Lorenzen J, Kuhl C, Sondermann E, et al. Imaging anisotropic and viscous properties of breast tissue by magnetic resonance-elastography. Magn Reson Med. 2005; 53(2):372–87.

53. Maderwald S, Uffmann K, Galban CJ, de Greiff A, Ladd ME. Accelerating MR elastography: a multi-echo phase-contrast gradient-echo sequence. J Magn Reson Imaging. 2006;23(5):774–80.

54. Bieri O, Maderwald S, Ladd ME, Scheffler K. Balanced alternating steady-state elastography. Magn Reson Med. 2006;55(2):233–41.

55. Huwart L, Salameh N, ter Beek L, Vicaut E, Peeters F, Sinkus R, et al. MR elastography of liver fibrosis: preliminary results comparing spin-echo and echo-planar imaging. Eur Radiol. 2008;18(11):2535–41.

56. Mariappan YK, Glaser KJ, Hubmayr RD, Manduca A, Ehman RL, McGee KP. MR elastography of human lung parenchyma: technical development, theoretical modeling and in vivo validation. J Magn Reson Imaging. 2011;33(6):1351–61 [Research Support, N.I.H., Extramural].

57. Mariappan YK, Kolipaka A, Manduca A, Hubmayr RD, Ehman RL, Araoz P, et al. Magnetic resonance elastography of the lung parenchyma in an in situ porcine model with a noninvasive mechanical driver: correlation of shear stiffness with trans-respiratory system pressures. Magn Reson Med. 2012;67(1):210–7 [Research Support, N.I.H., Extramural Research Support, Non-U.S. Gov't].

58. Aletras AH, Ding S, Balaban RS, Wen H. DENSE: displacement encoding with stimulated echoes in cardiac functional MRI. J Magn Reson. 1999;137(1): 247–52 [Comparative Study].

59. Robert B, Sinkus R, Gennisson JL, Fink M, editors. Application of DENSE MR-Elastography to the human heart: first in vivo results. In: The 17th annual conference of the international society for magnetic resonance in medicine; 2009; Honolulu, Hawaii, USA.

60. Yin M, Rouviere O, Glaser KJ, Ehman RL. Diffraction-biased shear wave fields generated with longitudinal magnetic resonance elastography drivers. Magn Reson Imaging. 2008;26(6):770–80.

61. Papazoglou S, Hamhaber U, Braun J, Sack I. Algebraic Helmholtz inversion in planar magnetic resonance elastography. Phys Med Biol. 2008;53(12):3147–58.

62. Yin M, Glaser JK, Talwalkar JA, Manduca A, Ehman RL, editors. Validity of a 2-D wave field model in MR elastography of the liver. In: Proceedings of the international society for magnetic resonance in medicine; 2009; Honolulu, Hawaii, USA.

63. Manduca A, Lake DS, Kruse SA, Ehman RL. Spatio-temporal directional filtering for improved inversion of MR elastography images. Med Image Anal. 2003; 7:465–73.

64. Manduca A, Muthupillai R, Rossman PJ, Greenleaf JF, Ehman RL. Visualization of tissue elasticity by magnetic resonance elastography. In: Hones K, Kinikis R, editors. Visualization in biomedical computing. Berlin: Springer; 1996. p. 63–8.

65. Oliphant TE, Manduca A, Ehman RL, Greenleaf JF. Complex-valued stiffness reconstruction for magnetic resonance elastography by algebraic inversion of the differential equation. Magn Reson Med. 2001;45(2):299–310.

66. Van Houten EE, Miga MI, Weaver JB, Kennedy FE, Paulsen KD. Three-dimensional subzone-based reconstruction algorithm for MR elastography. Magn Reson Med. 2001;45(5):827–37.

67. Van Houten EE, Doyley MM, Kennedy FE, Paulsen KD, Weaver JB. A three-parameter mechanical property reconstruction method for MR-based elastic property imaging. IEEE Trans Med Imaging. 2005; 24(3):311–24.

68. Manduca A, Oliphant TE, Dresner MA, Mahowald JL, Kruse SA, Amromin E, et al. Magnetic resonance elastography: non-invasive mapping of tissue elasticity. Med Image Anal. 2001;5(4):237–54.

69. Dzyubak B, Glaser K, Yin M, Talwalkar J, Chen J, Manduca A, et al. Automated liver stiffness measurements with magnetic resonance elastography. J Magn Reson Imaging. 2013;38(2):371–9 [Research Support, N.I.H., Extramural].

70. Yin M, Talwalkar JA, Venkatesh SK, Ehman RL, editors. MR elastography of dynamic postprandial hepatic stiffness augmentation in chronic liver disease. In: Proceedings of the International Society for Magnetic Resonance in Medicine; 2009; Honolulu, Hawaii, USA.

71. Klatt D, Hamhaber U, Asbach P, Braun J, Sack I. Noninvasive assessment of the rheological behavior of human organs using multifrequency MR elastography: a study of brain and liver viscoelasticity. Phys Med Biol. 2007;52:7281–94.

72. Huwart L, Sempoux C, Salameh N, Jamart J, Annet L, Sinkus R, et al. Liver fibrosis: noninvasive assessment with MR elastography versus aspartate aminotransferase-to-platelet ratio index. Radiology. 2007; 245(2):458–66.

Clinical Applications of Liver Magnetic Resonance Elastography: Chronic Liver Disease

4

Sudhakar K. Venkatesh

Introduction

Magnetic resonance elastography (MRE) of liver was first applied for clinical use in 2007 at the Mayo Clinic, Rochester, MN, USA [1]. Since its introduction, MRE has proven to be useful in the evaluation of chronic liver diseases (CLD) and the technique is now available at more than 100 centers around the world.

Untreated CLD regardless of etiology results in liver fibrosis and if progressive leads to cirrhosis and organ failure. Liver fibrosis progression is an important prognostic factor in the management of CLD [2, 3]. Early detection and accurate staging of the liver fibrosis is important as effective treatment is now available that may prevent fibrosis progression and even result in regression of fibrosis [3, 4]. The traditional approach for diagnosis of liver fibrosis includes clinical examination, laboratory tests, conventional imaging, and liver biopsy. Clinical examination although very valuable in the later stages of the disease is not a sensitive method to detect early fibrosis.

Liver biopsy (LB) is considered the gold standard for liver fibrosis. However, it is invasive and associated with post procedure pain and discomfort in nearly one-third of patients [5]. Biopsy is associated with complications such as significant hemorrhage in 0.3 % and carries a 0.01 % risk of death from the procedure [6]. All these risk factors may result in refusal of LB by patients [7, 8]. Approximately 1/50,000 part of liver is obtained with LB for histological evaluation and therefore prone for sampling errors [9, 10]. In addition, histological staging of the liver fibrosis is qualitative with high intra and inter-observer variability [9, 11, 12]. Therefore LB is not an ideal method to evaluate liver fibrosis especially if multiple serial evaluations are to be performed. However LB is still useful for confirmation of diagnosis, assessment of severity of necroinflammation, and to evaluate possible coexistent disease processes which are not possible with currently available non-invasive techniques.

Non-invasive techniques for assessment of liver fibrosis are preferred over LB. Serum markers and composite scores although easy to perform are not accurate enough for differentiating early stages of fibrosis [13–15]. Similarly, elevated serum liver enzymes and other laboratory tests for liver function are not sensitive to detect early fibrosis [16]. Conventional imaging techniques including ultrasound (US), computed tomography (CT), and magnetic resonance imaging (MRI) can demonstrate changes in liver surface, alterations in liver parenchyma texture, and volumetric changes that are characteristic of cirrhosis. However these signs are not sensitive and are usually seen only in advanced fibrosis

S.K. Venkatesh, M.D., F.R.C.R. (✉)
Department of Radiology, Mayo Clinic College of Medicine, Mayo Clinic, 200 First Street SW, Rochester, MN, USA
e-mail: venkatesh.sudhakar@mayo.edu

S.K. Venkatesh and R.L. Ehman (eds.), *Magnetic Resonance Elastography*,
DOI 10.1007/978-1-4939-1575-0_4, © Springer Science+Business Media New York 2014

and cirrhosis and therefore not reliable for diagnosis of early or mild fibrosis [17–19].

Elastography techniques are now available for assessment of CLD. Transient elastography (TE) is the most widely used ultrasound based elastography technique. However TE has a high failure rate especially in obese subjects. Experience with other ultrasound based elastography techniques is still evolving.

MRE is probably the most accurate non-invasive technique available for detection and staging of liver fibrosis [20–25] and performs better than serum tests [22, 24, 25] and TE [22]. MRE is reliable with excellent repeatability, reproducibility, and near perfect inter-observer agreement. Therefore, MRE can serve as a powerful non-invasive clinical tool for detection

and stratification of liver fibrosis, and potentially replace liver biopsy. In this chapter, clinical application of MRE in the evaluation of CLD and liver fibrosis will be described.

Performing a Clinical Liver MRE

MRE of the liver can be performed in most clinical subjects including those who are obese subjects (Fig. 4.1), transplant recipients, as well as in children [26]. MRE can be performed on both 1.5T and 3T clinical scanners. No special preparation is required except for 4–6 h fasting similar to a typical liver MRI study. During follow-up assessment it is important to perform MRE under fasting condition for valid comparisons. MRE may be

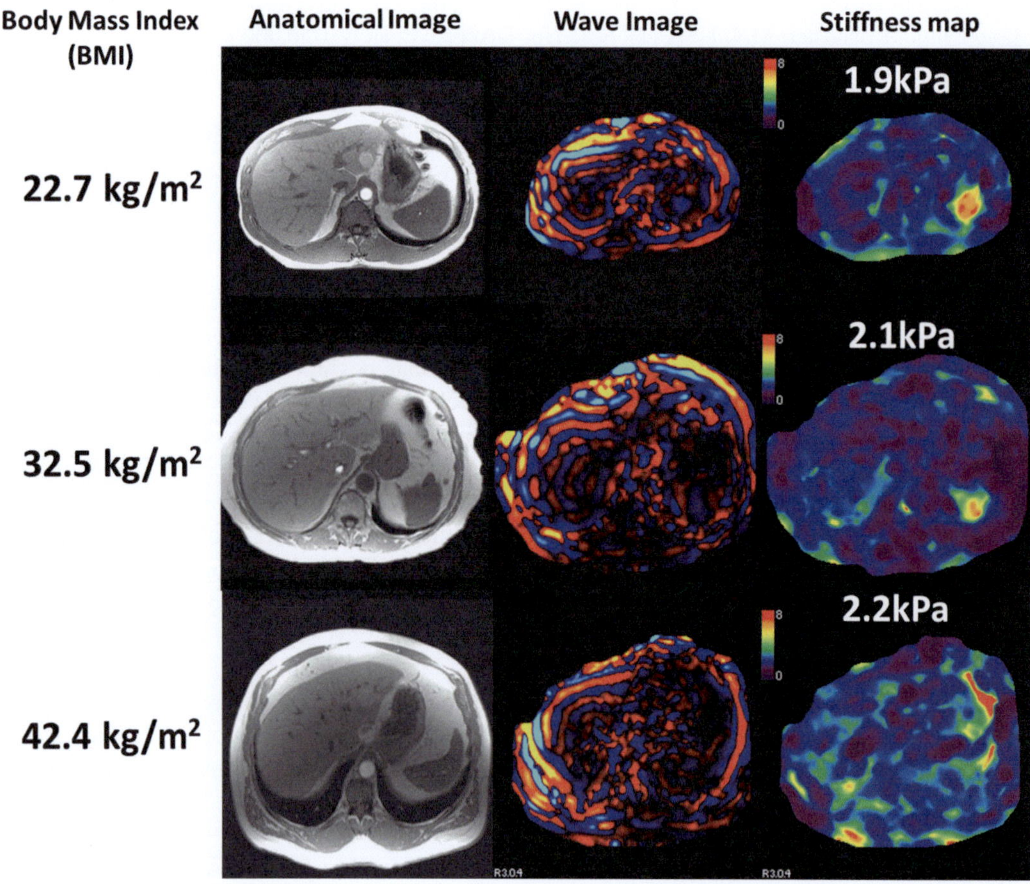

Fig. 4.1 MRE of liver can be performed in most clinical patients and of variable size. Anatomical MR image, wave image, and stiffness maps of three subjects (*each row*) with different body mass indices and normal liver. Note excellent propagation of shear waves in the liver in all three subjects and similar stiffness values of the livers

performed before or after administration of intravenous contrast agents such as gadolinium-DTPA and Gd-EOB-DTPA (Gadoxetate) without any significant influence on measured stiffness [27, 28]. MRE derived liver stiffness has excellent reproducibility and repeatability [29, 30]. The most common clinical technique is to perform MRE of liver at 60 Hz with four 10 mm slices through the largest cross-section of the liver. For more details on technique please see Chap. 3.

Measurement of Liver Stiffness

Stiffness of the liver can be quantified by drawing regions of interest (ROI) on the stiffness maps generated automatically on the scanner. Stiffness maps show spatial distribution of tissue stiffness across the cross-section. When drawing ROI, care must be taken to avoid large vessels, liver edge, fissures, gall bladder fossa, and regions of wave interference to obtain reliable stiffness values. Larger geographical ROI drawn over multiple slices is encouraged to ensure more volume of liver being sampled. The mean stiffness is reported in kilopascals (kPa). Liver stiffness with MRE is reproducible with high inter-observer agreement (>95 %) [29, 31–33]. Recently a study showed that the inter-observer agreement for MRE readers for liver stiffness was significantly better than that between pathologists staging liver fibrosis (0.99 vs. 0.91, $p < 0.001$) [34].

Normal Liver

Normal liver parenchyma (Fig. 4.2) is composed of hepatocytes, sinusoidal cells (endothelial cells, Kupffer cells, and other cells in space of Disse), sinusoidal spaces, and extracellular space of Disse. Hepatocytes account for nearly 65 % of the cells in the liver [35, 36]. The liver parenchyma is enclosed in a thin connective tissue layer (Glisson's capsule). The extracellular matrix (ECM) provides the architectural support and necessary scaffolding and is predominantly present in the space of Disse as basement membrane and around the portal areas, large vessels, and the capsule as interstitial matrix (connective tissue stroma). The ECM is composed mainly of collagen, glycoproteins, and proteoglycans [37]. The collagen content (mainly collagen Type I and III) in normal liver is very little, usually <1 % and is approximately 2–8 mg/g wet weight whereas in fibrotic liver this increases to several fold to the order of 30 mg/g wet weight or more [38, 39]. The blood in the sinusoids and vessels accounts for nearly 25–30 % of liver volume [40]. Liver has a dual blood supply: 70–80 % from the portal vein and 20–30 % from the hepatic artery. The sinusoids are supplied by portal venules and drain into central venules and hepatic veins into the inferior vena cava. The pressures in normal portal vein, portal venules, central venules, hepatic veins, and inferior vena cava are 5–10,

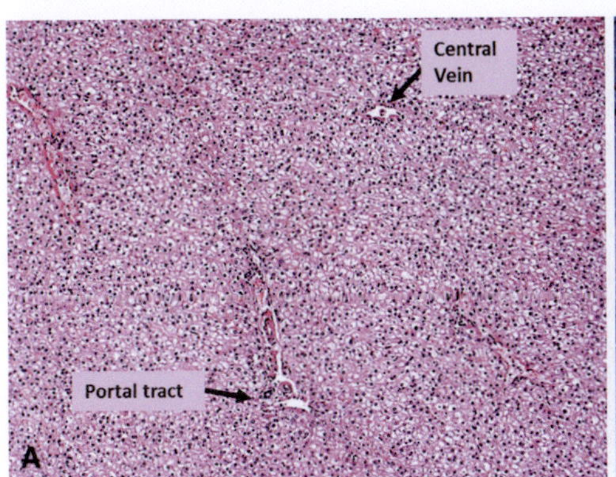

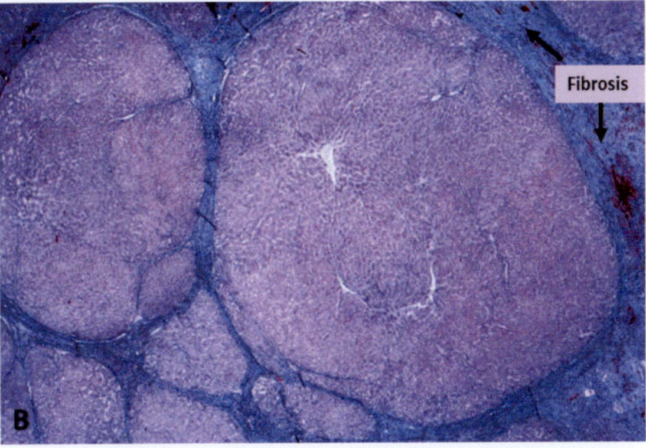

Fig. 4.2 Histology of normal liver (**a**) and cirrhotic liver (**b**) with Masson-Trichrome stain to highlight fibrosis. The fibrotic tissue (collagen) is *stained blue*. Normal liver has minimal fibrotic tissue around the portal tracts. Cirrhosis is characterized by multiple nodules of varying size encapsulated with fibrotic bands of variable size. (Images provided by Professor Aileen Wee, Pathology, National University of Singapore)

4–5, 1–3, <5, and −5 to 5 mmHg respectively [41–43]. The arterial blood enters the sinusoids at 9–18 mmHg pressure [44]. Blood flow through the liver is high volume (1–1.5 L/min) across a low resistance system due to large cross-sectional area of the sinusoids resulting in a low pressure perfusion system [45–48]. Due to this, liver is extremely compliant. If the portal blood flow is doubled, the portal pressure in normal liver only increases by approximately 2 mmHg by passive decrease in resistance [47, 48]. Normal liver therefore is a soft organ as it has very little fibrous tissue and a low pressure blood circulation system.

Normal Liver Stiffness

Normal liver is soft and has a mean stiffness value of 2.05–2.44 kPa and ranges from 1.54 to 2.87 kPa [20, 49–53]. The range of normal liver stiffness reported by different research groups may be influenced by the age, sex, body mass index (BMI), diet, and ethnicity of the subject populations studied, however no definite correlations are established. A few early studies have demonstrated that there is no significant correlation between liver stiffness and age, sex and BMI [49, 50], however, need to be verified in a larger study population comprising of healthy normal subjects. Normal liver stiffness does not show any significant diurnal variation or affected by a meal challenge [29, 30, 54] as it can easily accommodate the increased post prandial portal flow as explained in previous section. Fatty change or simple steatosis without any inflammation also does not significantly influence measured stiffness and is similar to normal liver [20, 49, 50].

CLD and Liver Fibrosis

CLD can result from many etiologies including viral hepatitis, excessive alcohol consumption, non-alcoholic steatohepatitis, drug toxicity, and chronic passive congestion due to right heart failure. Chronic injury to the liver parenchyma leads to necrosis and hepatocyte loss and heals by

fibrosis-laying of ECM. Liver fibrosis results in both quantitative and qualitative changes in the collagenous and non-collagenous components of the ECM [55]. The source of ECM changes in fibrotic liver can be hepatic stellate cells (HSC), fibroblasts and myofibroblasts of the portal tract, smooth muscle cells localized in vessel walls, and myofibroblasts localized around the centrolobular vein. Activation of these sources depends on etiology and initial site of liver injury. These changes involve portal tracts, central veins and sinusoids as well as hepatocytes. The perisinusoidal basement like matrix changes to interstitial type matrix containing fibril forming collagens. The collagen content of fibrotic liver is significantly higher than normal liver (three–tenfold or more). All of these contribute to mechanical and physical changes in the liver [55].

Different etiology produce different spatial patterns of fibrosis and very variable even among patients with same etiology [55]. For example, chronic viral hepatitis is characterized by interface hepatitis and development of portal to portal septa, portal-central septa, and septa ending blind in parenchyma; biliary cirrhosis tends to cause portal-portal bridging with preservation of central vein until later stages; non-alcoholic steatohepatitis (NASH), alcohol, and other metabolic diseases are characterized by perisinusoidal deposition of fibrillar matrix and around the hepatocytes known as chicken wire pattern [55, 56]. Typically CLDs result in inhomogeneous distribution of changes of inflammation and fibrosis.

Liver fibrosis is a dynamic process of accumulation of excessive fibrous tissue associated with degradation and remodeling in a balanced process [57]. When accumulation of fibrosis exceeds degradation, cirrhosis results sooner or later, characterized by formation of regenerative nodules encapsulated by fibrous tissue septa (Fig. 4.2) and associated changes in vascular architecture [56]. The progression of liver fibrosis in CLD is usually slow over several years and can increase rapidly in advanced fibrosis [58] and when immunosuppression coexists as in patients with liver transplants [59] or those with human immunodeficiency virus (HIV) co-infection [60].

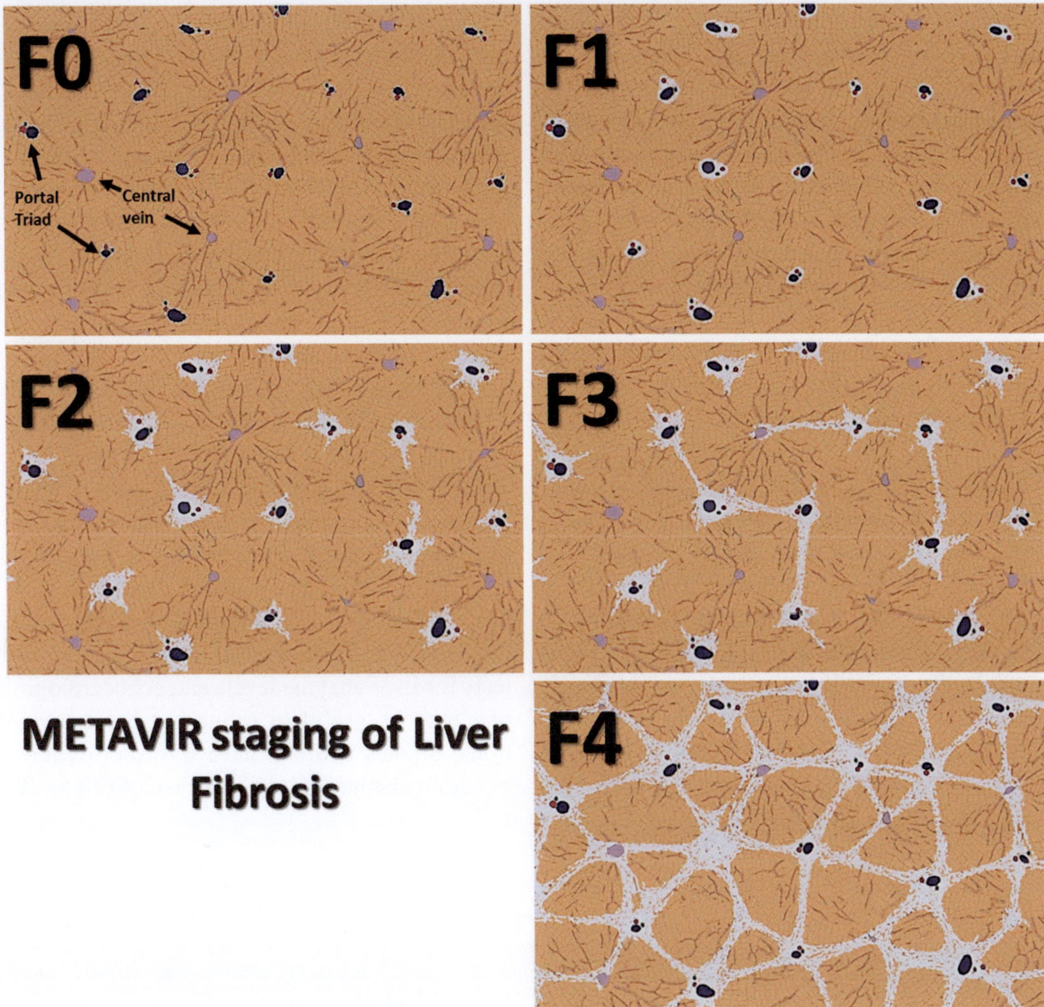

Fig. 4.3 Diagram illustrating METAVIR staging of liver fibrosis F0 through F4

Evaluation of progression of fibrosis has clinical implications for risk stratification, prognosis, and evaluation of therapy. The current gold standard of liver biopsy provides qualitative estimation of extent of fibrosis using ordinal scoring systems. Several fibrosis staging systems exist including Batts and Ludwig [61], Ishak [62] and METAVIR [63] for chronic hepatitis and Brunt Schema [64] for NASH. METAVIR system (Fig. 4.3) is widely used and can be briefly summarized as follows: F0-no fibrosis, F1-periportal fibrosis, F2-periportal fibrosis with rare septa, F3-numerous septa with cirrhosis, and F4-cirrhosis. Clinically significant fibrosis is defined as ≥F2

stage as it carries the highest risk of progression and cirrhosis in chronic hepatitis [65].

Histologically, the increasing severity of fibrosis or stages of fibrosis is based on identifying the location and amount of fibrosis, and assessment of qualitative features such as formation of septa, bridges, or nodules encapsulated by fibrous tissue. The staging systems provide information on relative severity of the process but are not accurate quantifiers of fibrosis content. Histological evaluation is however limited by sampling error and inter-observer variability.

Liver fibrosis content can also be assessed with quantification of liver fibrosis (collagen content) with histo-morphometric methods on liver biopsy samples using suitable stains like Sirius Red [66]. Fibrotic tissue highlighted by these techniques is expressed as percentage of total area of histological sample and is known as collagen proportionate area (CPA). Morphometric assessment can also be performed with digital methods using second harmonic generation (SHG) microscopy. With morphometric analysis, the amount of fibrous tissue ranges from 1 to 4 % in normal livers to 15 to 35 % in cirrhosis [66]. However, these methods are limited by sampling error inherent to liver biopsy.

Stiffness of Fibrotic Liver

Fibrotic livers are stiffer than a normal liver due to increased ECM. Preliminary study has shown that MRE derived stiffness of the liver correlates significantly ($r=0.78$, $p<0.001$) with fibrosis amount in liver samples [67] (Fig. 4.4). The degree of fibrosis (fibrosis burden) is variable in CLD from different etiology [55, 68–71].

For example chronic hepatitis C typically causes micronodular cirrhosis with smaller nodules and thicker septa, i.e., more fibrosis [71] as compared to chronic hepatitis B which is associated with macronodular cirrhosis with less fibrosis at the same stage of fibrosis [68]. Liver stiffness which correlates with amount of fibrosis may therefore be different among patients with the same histological fibrosis stage from the same etiology and also different from patients with the same fibrosis stage of different etiology.

Other conditions that may be associated with an increase in liver stiffness are acute biliary obstruction, acute inflammation of the liver [72], and passive congestion resulting from cardiac failure or other causes of raised central venous pressure [73]. MRE for detection or staging of liver fibrosis should be avoided when these conditions are known to be present as they may result in elevated stiffness measurements. Laboratory tests for liver enzyme levels and echocardiography may be useful in ruling out these conditions. Studies with TE have shown correlation of serum alanine aminotransferase (ALT) levels,

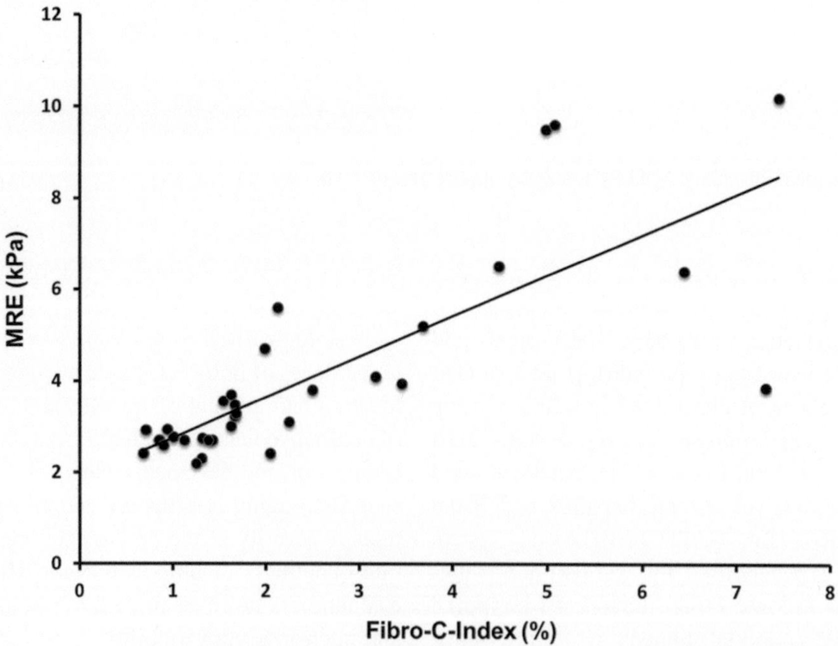

Fig. 4.4 Graph illustrating correlation between liver stiffness with MRE and fibrotic content with morphometric analysis (Fibro-C-Index) in liver biopsy of 32 patients with chronic hepatitis B. (Reproduced with permission from Venkatesh et al. [68])

a marker of liver inflammation with liver stiffness, however no such correlation has been shown with MRE [24].

Detection of Liver Fibrosis with MRE

MRE is accurate in distinguishing fibrotic livers from normal liver [20, 22, 23, 51, 74]. Early fibrosis can be detected even when anatomical features suggestive of fibrosis are absent (Fig 4.5) [17, 18]. MRE can differentiate normal liver from fibrotic livers with 89–99 % accuracy, 80–98 % sensitivity, and 90–100 % specificity [20, 21, 23, 24, 51, 75]. The cut-off stiffness values for detection of fibrosis range from 2.4 to 2.93 kPa in different studies [20, 21, 23, 24, 51, 75]. The different cut-off value is probably due to different populations studied and inclusion of multiple etiologies in the earlier studies. The degree, pattern, and distribution of fibrosis are different in CLD resulting from different etiologies as noted above. MRE provides a unique opportunity to observe stiffness distribution in the entire cross-section of the liver. Stiffness maps can show heterogeneity of liver stiffness that may reflect variations in the amount of fibrosis. Spatial distribution of stiffness in different CLDs may be observed (Fig. 4.6)

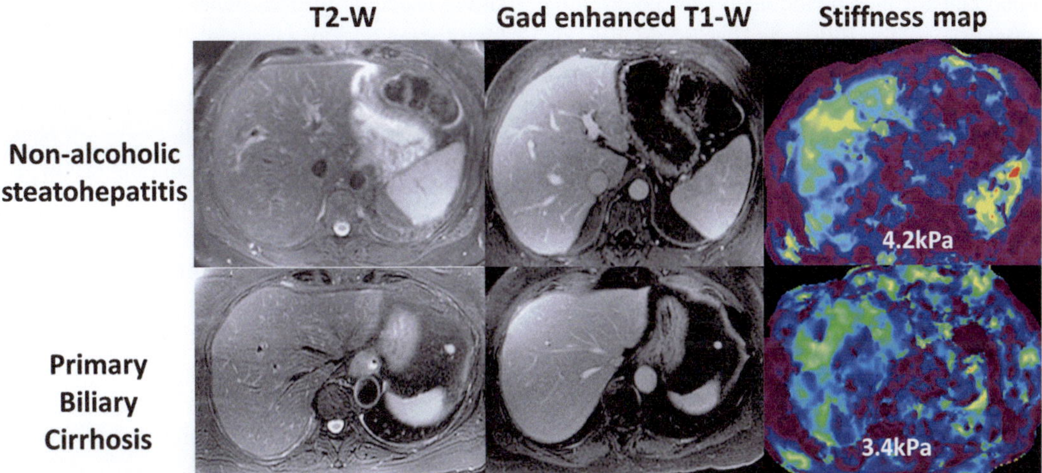

Fig. 4.5 MRE can detect increased stiffness due to fibrosis even when morphological appearance of liver is normal. Two examples of morphologically normal appearing liver but there was increased stiffness consistent with fibrosis and liver biopsy confirmed presence of significant fibrosis in both cases

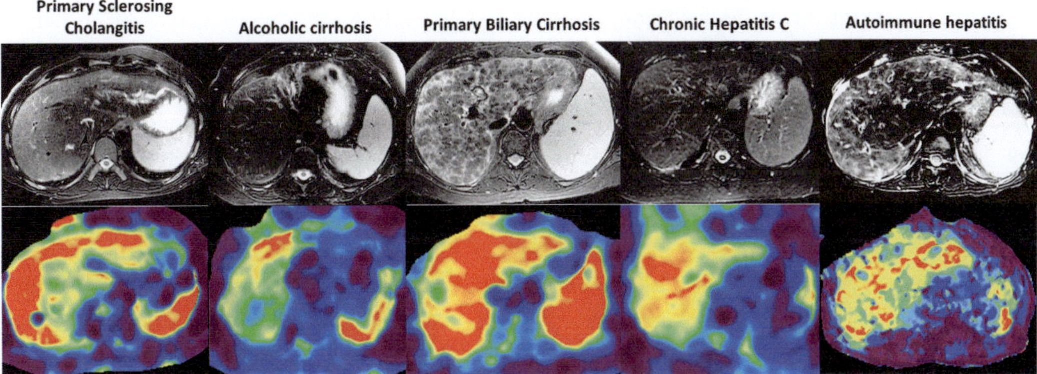

Fig. 4.6 Spatial distribution of stiffness in cirrhosis from different etiologies. Note the peripheral distribution of increased stiffness in primary sclerosing cholangitis and around the nodules in the case of autoimmune hepatitis

and this information may be useful for diagnosis as well as providing guidance for liver biopsy [76]. The influence of regional variation of fibrosis deposition on MRE measured stiffness is not known and needs to be evaluated in future studies. Recently MRE studies on single etiologies like viral hepatitis [24, 25], NAFLD [75, 77], Gaucher's [77], and alcoholic liver disease [78] are published and the cut-off values may be useful as references for these specific etiologies.

Staging of Liver Fibrosis with MRE

Accurate assessment of degree of fibrosis is important for therapeutic decisions, determining prognosis, and to follow up disease progression. The stiffness of the liver assessed with MRE increases with fibrosis stage (Fig. 4.7). Liver

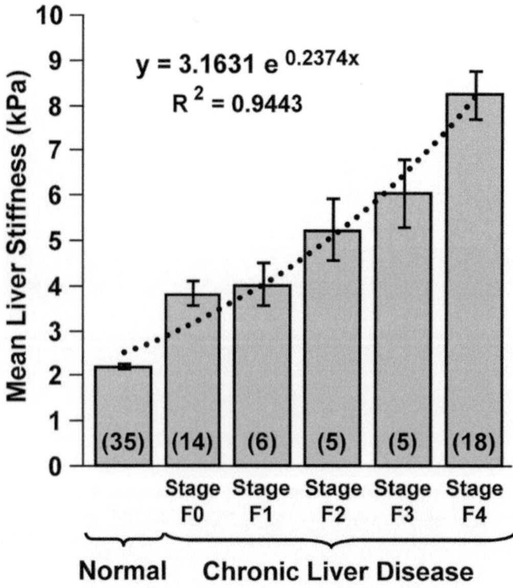

Fig. 4.7 Mean liver stiffness increases with increased fibrosis stage. Shown is a summary of the mean shear stiffness measurements of the liver for the 35 normal volunteers and the 48 patients divided into the five different fibrosis stages, which are indicated as F0, F1 … F4. Liver stiffness is significantly higher in patients than in the control group. The standard errors for each group are also illustrated in error bar for each group. An exponential function fit well to the liver stiffness data with an R^2 value of 0.94. (Reproduced with permission from Yin et al. [21]: http://www.elsevier.com)

stiffness shows excellent linear correlation with histologic fibrosis stages (spearman correlation coefficient $r=0.8–0.97$) [20, 23–25, 79]. MRE also correlates well with the amount of fibrosis assessed with morphometric analysis (Fig. 4.4) [67]. The trend of increased stiffness with increased stages of fibrosis has been demonstrated in the most common CLDs (Figs. 4.8, 4.9, and 4.10) by several researchers around the world.

MRE can differentiate individual stages of fibrosis with excellent accuracy. For clinical management purposes, it is useful to stage fibrosis into groups: mild fibrosis ($\geq$F1), clinically significant fibrosis ($\geq$F2), advanced fibrosis ($\geq$F3), and cirrhosis (F4). Good to excellent sensitivity, specificity, and accuracy of MRE in differentiating various stages of fibrosis have been shown in several studies (see Table 4.1). The cutoffs for significant fibrosis and advanced fibrosis are variable in different studies and probably related to the variability in the amount of fibrosis that occurs in different etiologies [24, 25, 78].

Liver stiffness may be affected by parenchymal inflammation which often accompanies chronic viral hepatitis although the effect may not be significant. Studies on chronic viral hepatitis have shown that inflammation does not affect correlation of stiffness with fibrosis stage [67, 79]. The confounding effect of inflammation will probably be most apparent in differentiating earlier stages of fibrosis from normal livers. More studies are required to establish the relationship between inflammation and liver stiffness. Similarly there is no correlation between fatty change and liver stiffness measured in livers (Fig. 4.11).

Clinically significant fibrosis ($\geq$F2) can be differentiated from mild fibrosis (F0–F1) with an excellent accuracy of >92 % and positive predictive value of >93 %. Similarly cirrhosis can be diagnosed with accuracy of >95 % and negative predictive value of >98 %. MRE can therefore accurately *rule in* significant fibrosis and *rule out* cirrhosis which is very useful for clinical decision making. MRE has lower accuracy for differentiation of mild fibrosis from normal and inflammation only (F0). This may result in misclassification of mild fibrosis as normal liver and vice versa, however, this should occur in a very

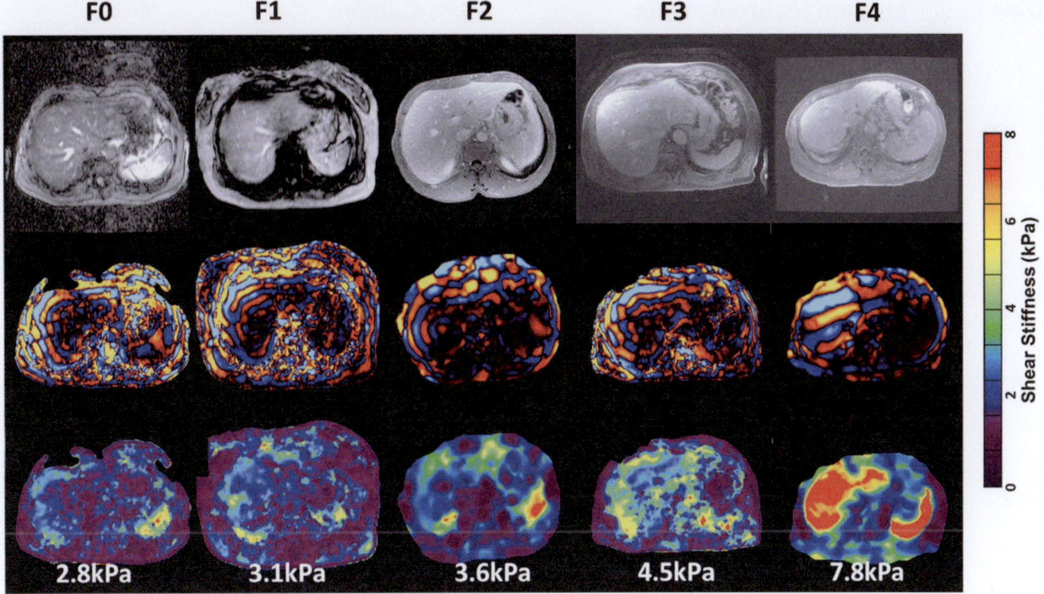

Fig. 4.8 Examples of histology confirmed cases of chronic hepatitis C with fibrosis stages F0–4. The *color bar* on the *right* is shear stiffness scale from 0 to 8 kPa

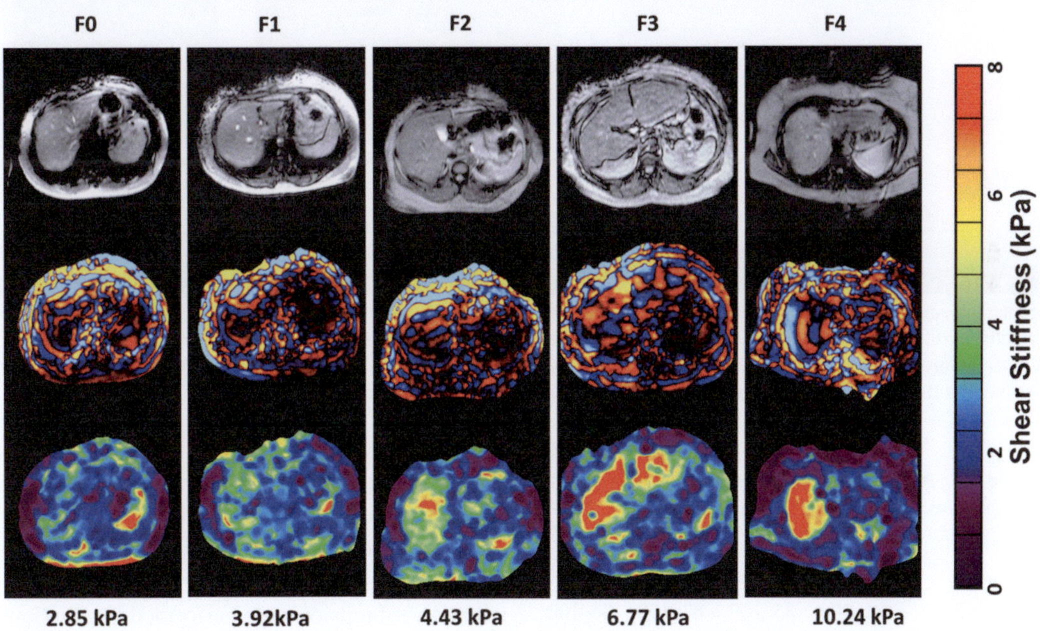

Fig. 4.9 Examples of increasing stiffness with increasing stage of fibrosis (F0-F4) with biopsy proven cases of non-alcoholic fatty liver disease (NAFLD). The *color bar* on the *right* is shear stiffness scale from 0 to 8 kilopascals (kPa)

small proportion of patients and usually only by one stage [23, 24] but this performance is better than TE [80]. The misclassification may be related to biopsy sampling and heterogeneity of liver fibrosis. Laboratory tests and clinical findings are useful for confirmation in such cases or a

Stage	Masson trichrome stain	Fibro-C-Index	MRE

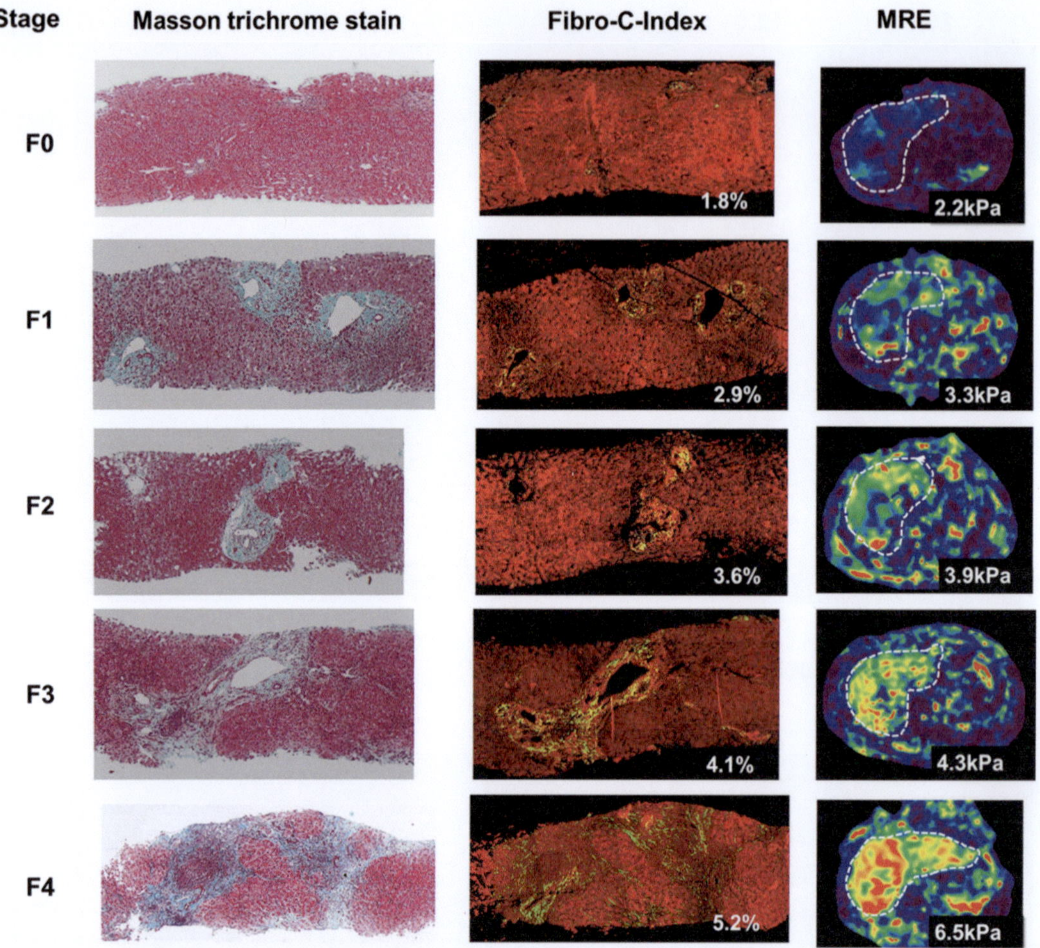

Fig. 4.10 Correlation of liver stiffness with increasing stages of fibrosis and morphometric assessment of fibrosis content in cases of chronic Hepatitis B. The *first column* is the Masson-Trichrome stained images of biopsy samples to highlight collagen. The *second column* shows results from Fibro-C-index, a morphometric method to estimate collagen content in biopsy samples and the *third column* shows stiffness maps with MRE in the same patient. (Reproduced with permission from Venkatesh et al. [68])

Table 4.1 Accuracy of MRE for staging liver fibrosis compiled from [20, 22–25, 51, 74]*

Stage	≥F1	≥F2	≥F3	≥F4
Sensitivity	72–98	86–100	78–100	89–100
Specificity	81–100	85–100	87–100	86–100
Accuracy	90–99	92–99	92.3–99	95–100
Positive predictive value	97–100	93–100	92–100	86–100
Negative predictive value	64–86	93–100	95–100	98–100

*numbers are percentages

repeat MRE study after short follow up may be helpful. Rarely a liver biopsy may be required to resolve discordant findings.

Proposed guidelines for interpretation of liver stiffness values are summarized in Table 4.2. The guidelines are based on our experience of more than 3,500 patients evaluated in clinical practice where liver biopsies were performed in a random sample of patients. The fibrosis stages are categorized as stages 1–2, 2–3, 3–4 as histologic staging is qualitative (categorical), whereas liver stiffness is a quantitative parameter on a continuous scale. Correlating the two scores on one-on-one basis is not statistically feasible but the grouping of stages into categories such as F1–2, F2–3 and F3–4, and F4 is done to facilitate clinical decision making.

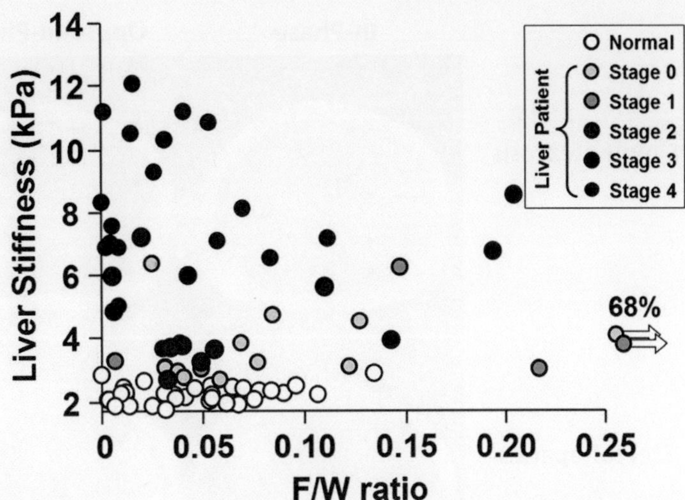

Fig. 4.11 Liver fat content does not influence stiffness with MRE. There is no correlation observed between fat/water (F/W) ratio and the liver stiffness measured in a group of 35 normal healthy volunteers and 48 patients ($R^2 < 0.05$) (Reproduced with permission from Yin et al. [21]: http://www.elsevier.com)

Table 4.2 Proposed guidelines for interpretation of liver stiffness with MRE performed at 60 Hz frequency

<2.5 kPa	Normal
2.5–2.9	Normal or inflammation
2.9–3.5	Stage 1–2 fibrosis
3.5–4	Stage 2–3 fibrosis
4–5	Stage 3–4 fibrosis
>5	Stage 4 fibrosis or cirrhosis

Cirrhosis was previously considered a terminal stage of CLD, but is now further classified into compensated and decompensated cirrhosis. Decompensated cirrhosis is characterized by the presence of variceal bleeding, ascites, or hepatic encephalopathy and carries several fold increase risk of mortality than compensated cirrhosis. Decompensation may be related to overall fibrosis burden. Increased liver stiffness is related to increased risk of decompensated cirrhosis, development of HCC, and mortality [81]. In a recent study, liver stiffness measured with MRE was independently associated with decompensation. A stiffness value of ≥5.8 kPa was associated with higher risk of decompensation [82].

MRE in Non-alcoholic Fatty Liver Disease

Non-alcoholic fatty liver disease (NAFLD) is the most common cause of liver disorder in Western countries [83]. The spectrum of NAFLD consists of NAFL or simple steatosis (SS), NASH, and NASH with fibrosis. NASH and NASH related fibrosis is associated with significantly higher liver related mortality (~15 %) as compared to those with SS alone (<2.5 %) [84–86]. It is important to distinguish these two conditions for clinical management. Simple steatosis progresses to NASH in 10–20 % of cases, 10–15 % of NASH progresses to cirrhosis, and nearly 25–30 % of advanced fibrosis progresses to cirrhosis [86, 87]. Conventional imaging modalities are not able to reliably differentiate the two entities. MRE is useful (Fig. 4.12) as SS does not affect liver stiffness [20, 24, 88] whereas both NASH and fibrosis demonstrate an increase in liver stiffness [75, 89].

In a preliminary study with 58 NAFLD patients, Chen et al. [75] demonstrated that SS can be differentiated from NASH ± fibrosis, with 94 % sensitivity, 73 % specificity, and an accuracy of 0.93 using a cut-off value of 2.74 kPa. The mean hepatic stiffness of patients with SS was significantly less than those with steatohepatitis (2.51 vs. 3.24 kPa, $p = 0.028$). The mean liver stiffness of patients with steatohepatitis but no fibrosis was significantly less than that for patients with hepatic fibrosis (4.16 kPa, $p = .030$).

In an another study, Kim et al. [89] showed that MRE with a cut-off value of 4.15 kPa is useful in detecting advanced fibrosis (F3–F4) in NAFLD with 85 % sensitivity, 93 % specificity, and an accuracy of 0.954. MRE is therefore a promising technique for differentiating SS from

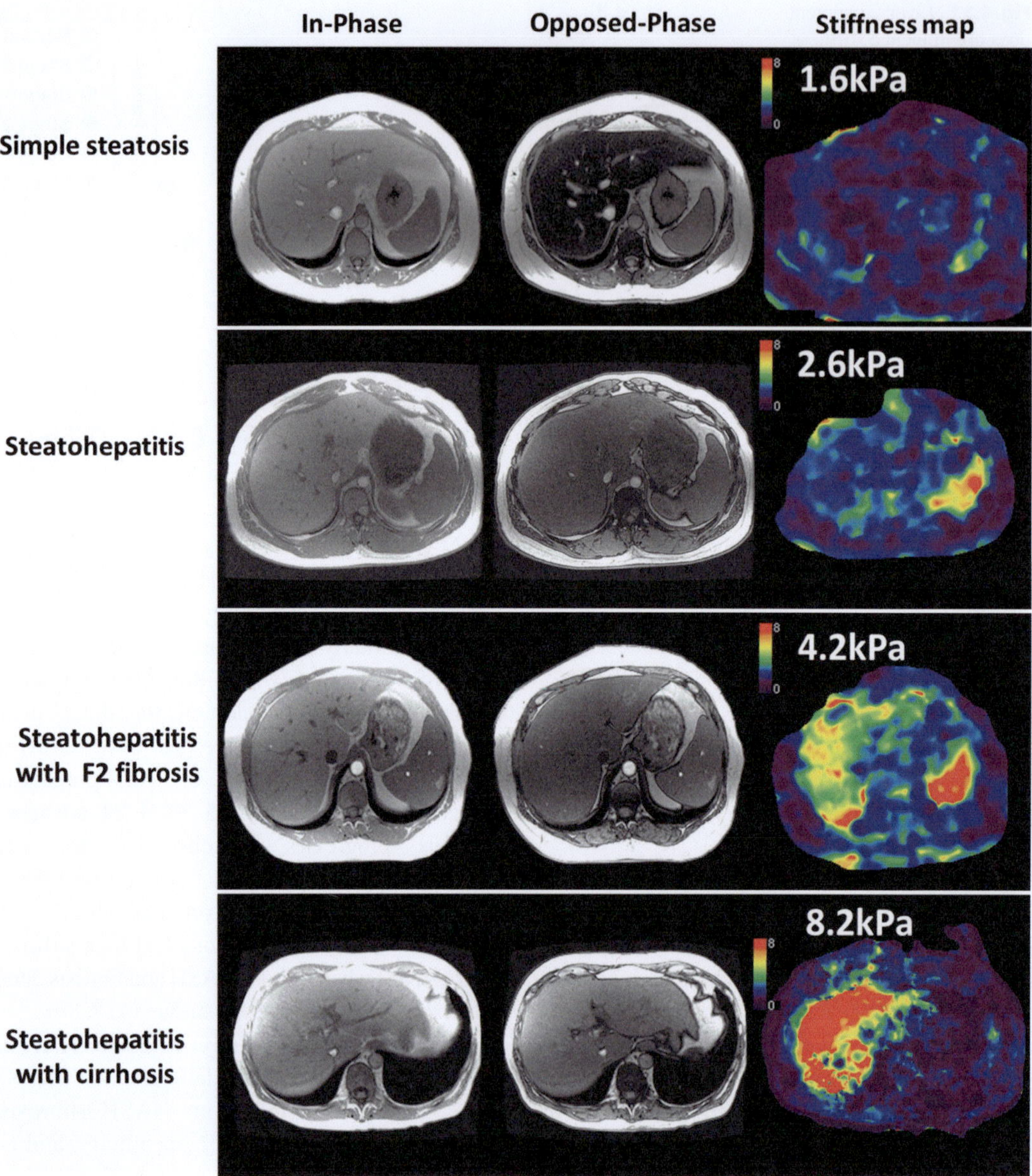

Fig. 4.12 MRE in NAFLD. Case examples of histology confirmed simple steatosis (*top row*), steatohepatitis only (*second row*), steatohepatitis with fibrosis (*third row*), and steatohepatitis with cirrhosis (*bottom row*). The in- and opposed phase gradient echo images showing signal loss characteristic of fatty livers most prominent in simple steatosis and steatohepatitis cases. Note steatohepatitis results in a mild increase in liver stiffness whereas fibrosis and cirrhosis are associated with significant increase in liver stiffness

NASH as well as stratifying NASH patients with advanced fibrosis. MRE will probably have a significant role in NAFLD as obesity-related CLD is a major health burden in USA. A liver MRI study including fat quantification techniques and MRE can serve as a comprehensive one-stop technique evaluation of NAFLD.

Clinical Follow Up with MRE

Prognosis and management of CLD depends on the amount of liver fibrosis and its progression [58]. Clinical trials of chronic viral hepatitis patients have demonstrated that hepatic collagen

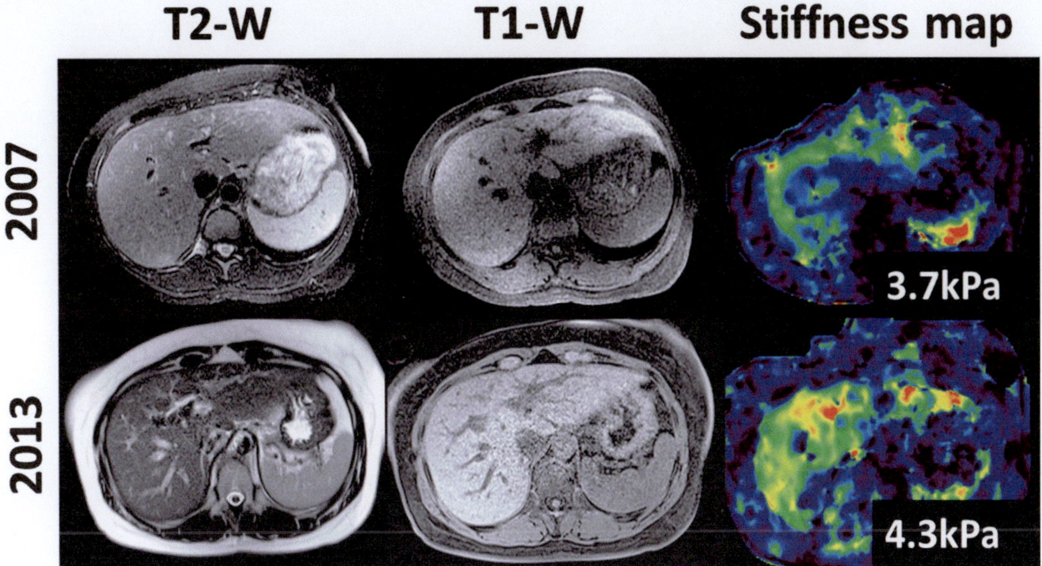

Fig. 4.13 MRE in longitudinal clinical follow up. A chronic hepatitis C patient at initial presentation in 2007 had elevated stiffness that correlated with F2 stage fibrosis with liver biopsy. Liver stiffness increased to 4.3 kPa after 5 years and suggestive of progression of fibrosis to at least F3 stage fibrosis

content on average doubles every 2 years and therefore patients with longer duration of disease are likely to have more fibrosis [58, 69, 90, 91]. The morbidity and mortality of patients with CLD increases once cirrhosis develops. Same etiopathology process at the same fibrosis stage is likely to produce different amounts of collagen in different individual patients [66]. Histological staging is not sensitive enough to measure minor changes in the fibrosis amount and fibrosis progression through histological stages does not always represent an increase in fibrosis. It is important to monitor fibrosis over time as the progression of fibrosis is variable. MRE may be a useful surrogate marker for liver fibrosis burden [67] and monitoring with MRE is possible. Clinical experience has shown that MRE can demonstrate changes in stiffness over time in patients on clinical follow up (Fig. 4.13).

Assessment of Treatment Response

Treatment for several CLDs is now available and studies have demonstrated regression of fibrosis in patients undergoing effective treatment [58, 92–96]. Increasing evidence for reversibility of liver fibrosis [97, 98] makes it necessary to monitor fibrosis and not merely make a diagnosis of liver fibrosis. Unfortunately a significant number of subjects on treatment show progression of disease [58]. MRE may be able to demonstrate the changes in the liver stiffness secondary to changes in the liver collagen content. Our preliminary clinical experience (Figs. 4.14, 4.15, and 4.16) has demonstrated the utility of MRE in assessing the changes in the liver stiffness which may be a useful biomarker for liver fibrosis content. Larger studies demonstrating the use of MRE in the assessment of treatment response are awaited.

MRE in Other Diffuse Liver Diseases

MRE is useful in the assessment of fibrosis resulting from use of hepatotoxic drugs such as methotrexate (Fig. 4.17). Preliminary study has shown that MRE is useful in detection of fibrosis due to methotrexate toxicity [99]. CLD can also result from chronic congestive hepatopathy resulting from right heart failure or secondary

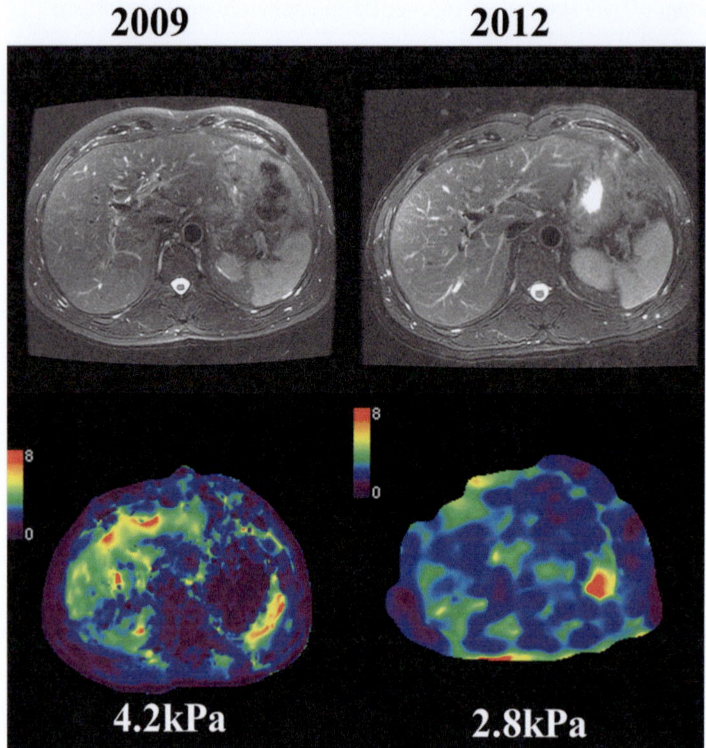

Fig. 4.14 MRE for assessment of treatment response in a patient with chronic hepatitis C. In 2009 patient received antiviral treatment and the baseline liver stiffness was 4.2 kPa which reduced to 2.8 kPa after 3 years. Interestingly both the liver function tests were normal at baseline and follow up

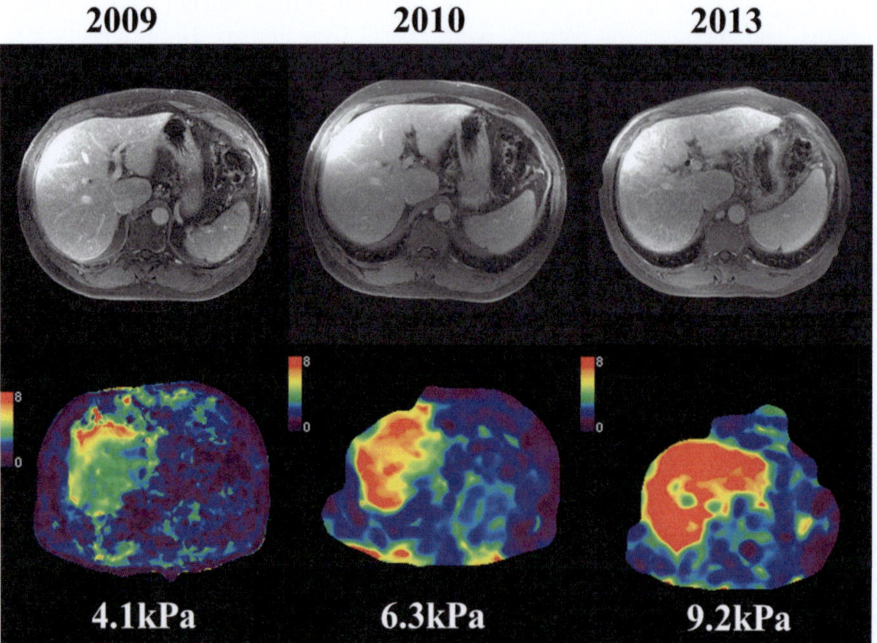

Fig. 4.15 MRE for assessment of treatment response. A patient with chronic hepatitis C and a baseline liver stiffness of 4.1 kPa was treated with antiviral therapy. The liver stiffness increased to 6.3 kPa a year later while on therapy. The antiviral therapy was changed but the liver stiffness increased significantly to 9.2 kPa after 3 years of treatment. Note the prominent caudate lobe after 1 year and volumetric changes most apparent at last follow up

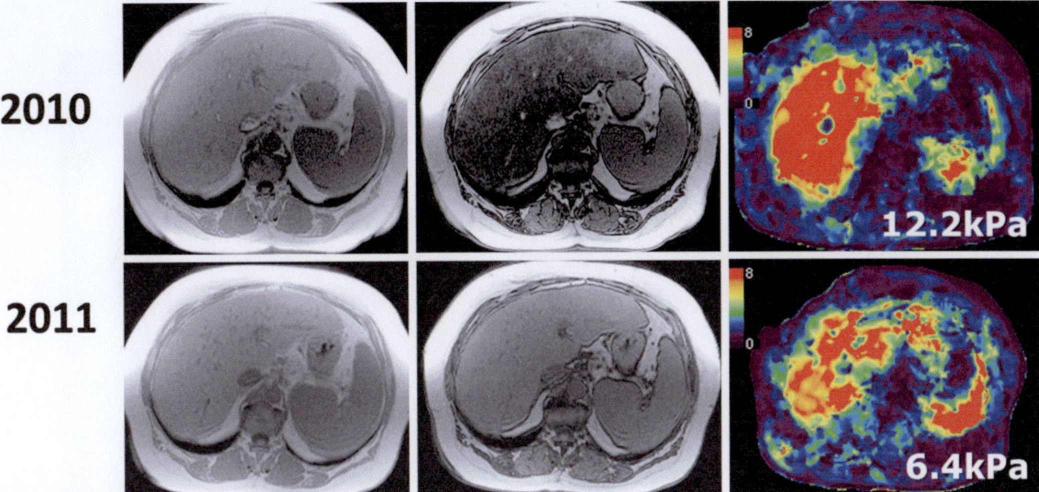

Fig. 4.16 MRE for assessment of treatment response. NASH. Obese patient with baseline MRI showed 25 % fat fraction and a very stiff liver due to steatohepatitis. Follow up MRI 1 year later and about 8-pound weight loss, the fat fraction was only 6 % and liver stiffness significantly improved

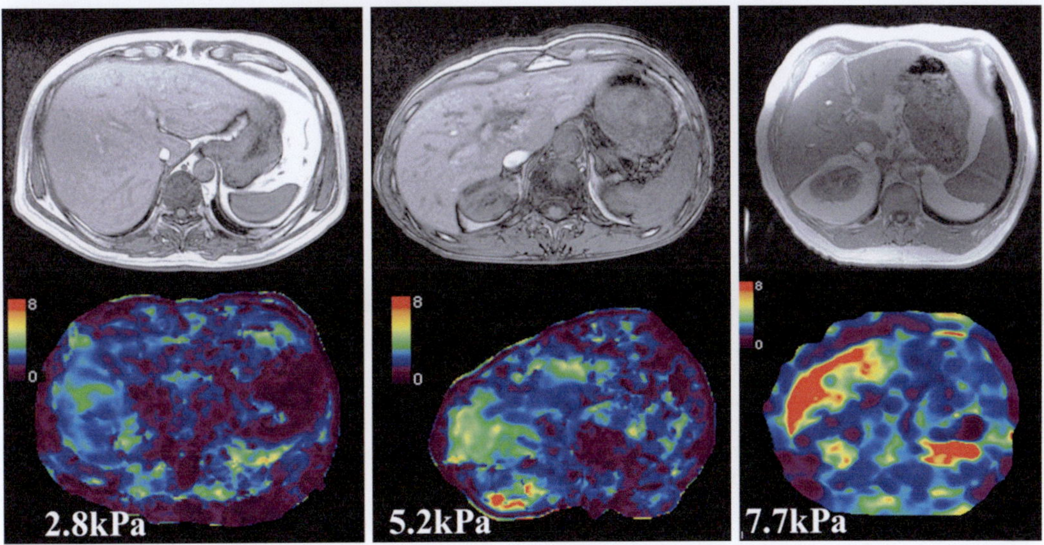

Fig. 4.17 MRE in the assessment of patients receiving methotrexate. Three patients with psoriasis received a total cumulative dose of 2.6 g (*first column*), 4 g (*second column*), and 4 g (*third column*) of methotrexate respectively. MRE shows significant liver stiffness in patients who received 4 g dose consistent with advanced fibrosis and cirrhosis. Note that the third patient is also obese which was thought to be a second risk factor for liver fibrosis

to congenital heart disease [100]. Studies have demonstrated the utility of MRE in the assessment of fibrosis in transplant livers [79, 101, 102]. Several other applications continue to emerge.

Limitations of MRE

The most widely used 2D-GRE-MRE technique is a breath hold sequence and is susceptible to motion induced artifacts. It is now possible to

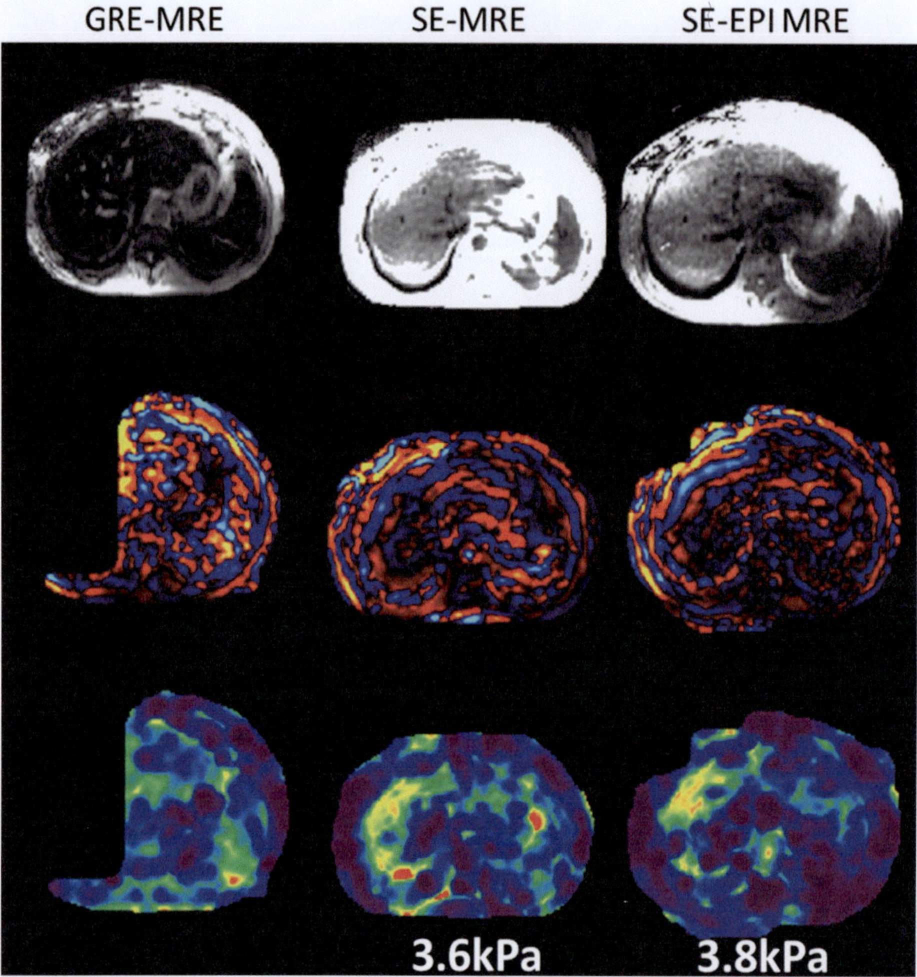

Fig. 4.18 Advanced MRE for evaluation of liver stiffness in iron over load livers illustrated here with a case of hemochromatosis and liver fibrosis. MRE performed with conventional 2D- GRE-MRE sequence and two iron overload sequences. There is low signal in the liver on the GRE-MRE sequence and therefore poor visualization of propagation of shear waves through the liver. In the spin-echo (SE-MRE) sequence and spin-echo echo-planar (SE EPI-MRE) sequence, the signal in liver is improved for visualization of shear wave propagation and reliable stiffness maps are produced. Note similar stiffness values obtained with both iron overload MRE sequences

obtain a single slice of MRE in as short a time as 11–12 s which will be useful to obtain good quality images. Improving acquisition time with reduced k-space acquisition and use of parallel imaging may help minimize breath hold artifacts. As the 2D-GRE-MRE sequence is sensitive to presence of iron in the liver, MRE in patients with moderate to severe iron overload due to hemochromatosis or hemosiderosis may fail as the signal from the iron loaded liver is too low such that shear waves cannot be visualized limiting the capability of inversion algorithm to process the stiffness map. This is probably the most common cause of failure of 2D-GRE-MRE in clinical practice. It should be noted however that it is due to low signal from the liver and not due to failure of propagation of waves. Implementing shorter TEs can improve the liver signal. Alternative pulse sequences with shorter TE such as spin-echo EPI sequence (Fig. 4.18) are capable of visualizing the waves through the liver [103]. In rare cases, the iron overload may be severe and it may be useful to perform MRE after treatment to reduce iron content in liver.

Comparison of MRE with Other Methods for Evaluation of Liver Fibrosis

The diagnostic performance of MRE is significantly higher than most common serum markers for liver fibrosis [22–25]. Although there is good correlation between MRE and TE [104, 105], MRE performs better than TE for detection and staging of liver fibrosis [23] with higher technical success. In a preliminary study for cross-validation of MRE and shear wave elastography (SWE), both MRE and SWE had similar success rates (96.4 vs. 92.2 %, $p=0.17$) [106]. However, SWE showed more unreliable results than MRE (13.1 vs. 0 %, $p<0.05$). Acoustic radiation force impulse (ARFI) imaging is another promising technique; however there are no reported studies comparing ARFI and MRE. Recently a meta-analysis of MRE and ARFI studies showed that MRE is more accurate than ARFI for detection of early stages of fibrosis and similar performance for advanced fibrosis [107].

Studies have compared MRE with routine qualitative analysis of morphological features of fibrosis and cirrhosis for detection of fibrosis and advanced fibrosis [17, 18]. MRE was clearly superior to detection of anatomical features of fibrosis/cirrhosis on MRI. Diffusion weighted imaging (DWI) has been studied by various investigators for evaluation of liver fibrosis. Comparison of MRE with DWI has shown that MRE is superior to DWI for detection and staging of liver fibrosis [108, 109]. A recent meta-analysis that investigated DWI and MRE for hepatic fibrosis showed that MRE performs much better than DWI [110]. In a comparison study, Choi et al. demonstrated that MRE is superior to Gadoxetate disodium enhanced MRI for detection and staging of liver fibrosis. MRE more strongly correlated with fibrosis staging than Gadoxetate enhanced MRI [111]. In an another interesting study, Motosugi et al. found that liver stiffness with MRE predicted insufficient liver enhancement with Gadoxetate sodium in patients with chronic hepatitis C and Child-Pugh Class A disease [112]. These results are probably reflec-tive of reduced liver function secondary to liver fibrosis. 31P-MR spectroscopy has been studied with interest with mixed success for correlation with liver fibrosis stage [113–116]. In a comparison study, Godfrey et al. found that MRS findings did not correlate with histologic fibrosis stage whereas MRE had high performance and corre-lated with fibrosis stage [32].

Future of Liver MRE

The liver MRE technique is undergoing several modifications: reducing scan time, improving resolution and increasing signal intensity, and so forth. One of the important areas to focus is to differentiate pathological processes such as inflammation, edema, and passive congestion from liver fibrosis by using additional parameters such as anisotropy, wave attenuation, and working with different mechanical models. Entire liver volume (volumetric) assessment is possible with 3-dimensional (3D) MRE sequence which may provide an opportunity to assess overall liver fibrosis burden, a parameter that could predict the final outcome in CLD. 3D acquisitions are also promising for characterization of focal lesions. Many research opportunities therefore seem apparent. MRE may play a greater role in evaluation of progression or regression of liver fibrosis and serve as a reference standard for assessment of performance of treatment methods and clinical management.

Liver MRE is an exciting and novel imaging technique with proven clinical value in the assessment of CLD. The clinical indications continue to emerge and technical improvements are ongoing for a promising future.

References

1. Venkatesh SK, Yin M, Talwalkar JA, Ehman RL, editors. Application of liver MR elastography in clinical practice. In: Proceedings of 17th annual meeting of ISMRM, Toronto; 2008 (abstract 2611).
2. Chu CM, Liaw YF. Hepatitis B virus-related cirrhosis: natural history and treatment. Semin Liver Dis. 2006;26(2):142–52.

3. Liver EAFTSOT. EASL clinical practice guidelines: management of chronic hepatitis B virus infection. J Hepatol. 2012;57(1):167–85.

4. Ghany MG, Nelson DR, Strader DB, Thomas DL, Seeff LB. Diseases AAfSoL. An update on treatment of genotype 1 chronic hepatitis C virus infection: 2011 practice guideline by the American Association for the Study of Liver Diseases. Hepatology. 2011;54(4)):1433–44.

5. Castéra L, Nègre I, Samii K, Buffet C. Pain experienced during percutaneous liver biopsy. Hepatology. 1999;30(6):1529–30.

6. Piccinino F, Sagnelli E, Pasquale G, Giusti G. Complications following percutaneous liver biopsy. A multicentre retrospective study on 68,276 biopsies. J Hepatol. 1986;2(2):165–73.

7. Bravo AA, Sheth SG, Chopra S. Liver biopsy. N Engl J Med. 2001;344(7):495–500.

8. Cadranel JF, Rufat P, Degos F. Practices of liver biopsy in France: results of a prospective nationwide survey. For the group of epidemiology of the French Association for the Study of the Liver (AFEF). Hepatology. 2000;32(3):477–81.

9. Regev A, Berho M, Jeffers L, Milikowski C, Molina E, Pyrsopoulos N, et al. Sampling error and intraobserver variation in liver biopsy in patients with chronic HCV infection. Am J Gastroenterol. 2002; 97(10):2614–8.

10. Maharaj B, Maharaj R, Leary W, Cooppan R, Naran A, Pirie D, et al. Sampling variability and its influence on the diagnostic yield of percutaneous needle biopsy of the liver. Lancet. 1986;1(8480):523–5.

11. Bedossa P, Dargere D, Paradis V. Sampling variability of liver fibrosis in chronic hepatitis C. Hepatology. 2003;38(6):1449–57.

12. Rousselet MC, Michalak S, Dupré F, Croué A, Bedossa P, Saint-André JP, et al. Sources of variability in histological scoring of chronic viral hepatitis. Hepatology. 2005;41(2):257–64.

13. Lin ZH, Xin YN, Dong QJ, Wang Q, Jiang XJ, Zhan SH, Sun Y, Xuan SY. Performance of the aspartate aminotrasnferase-to-platelet ratio index for the staging of hepatitis C-related fibrosis:an updated meta-analysis. hepatology 2011;53(3):726–36.

14. Halfon P, Bacq Y, De Muret A, Penaranda G, Bourliere M, Ouzan D, et al. Comparison of test performance profile for blood tests of liver fibrosis in chronic hepatitis C. J Hepatol. 2007;46(3):395–402.

15. Castera L. Noninvasive methods to assess liver disease in patients with hepatitis B or C. Gastroenterology. 2012;142(6):1293–302.e4.

16. Rockey DC, Bissell DM. Noninvasive measures of liver fibrosis. Hepatology. 2006;43(2 Suppl 1): S113–20.

17. Castera L. Invasive and non-invasive methods for the assessment of fibrosis and disease progression in chronic liver disease. Best Pract Res Clin Gastroenterol. 2011;25(2):291–303.

18. Venkatesh SK, Takahashi N, Glocker JF, Yin M, Talwalkar JA, Grimm RC, et al., editors. Non-invasive diagnosis of liver fibrosis: conventional MR imaging findings versus MR elastography. In: Proceedings of 17th annual meeting of ISMRM, Toronto; 2008 (abstract 2613); Toronto: ISMRM.

19. Rustogi R, Horowitz J, Harmath C, Wang Y, Chalian H, Ganger DR, et al. Accuracy of MR elastography and anatomic MR imaging features in the diagnosis of severe hepatic fibrosis and cirrhosis. J Magn Reson Imaging. 2012;35(6):1356–64.

20. Martí-Bonmatí L, Delgado F. MR imaging in liver cirrhosis: classical and new approaches. Insights Imaging. 2010;1(4):233–44.

21. Yin M, Talwalkar JA, Glaser KJ, Manduca A, Grimm RC, Rossman PJ, et al. Assessment of hepatic fibrosis with magnetic resonance elastography. Clin Gastroenterol Hepatol. 2007;5(10):1207–13.e2.

22. Huwart L, Peeters F, Sinkus R, Annet L, Salameh N, ter Beek LC, et al. Liver fibrosis: non-invasive assessment with MR elastography. NMR Biomed. 2006;19(2):173–9.

23. Huwart L, Sempoux C, Salameh N, Jamart J, Annet L, Sinkus R, et al. Liver fibrosis: noninvasive assessment with MR elastography versus aspartate aminotransferase-to-platelet ratio index. Radiology. 2007;245(2):458–66.

24. Huwart L, Sempoux C, Vicaut E, Salameh N, Annet L, Danse E, et al. Magnetic resonance elastography for the noninvasive staging of liver fibrosis. Gastroenterology. 2008;135(1):32–40.

25. Venkatesh SK, Wang G, Lim SG, Wee A. Magnetic resonance elastography for the detection and staging of liver fibrosis in chronic hepatitis B. Eur Radiol. 2014;24(1):70–8.

26. Ichikawa S, Motosugi U, Ichikawa T, Sano K, Morisaka H, Enomoto N, et al. Magnetic resonance elastography for staging liver fibrosis in chronic hepatitis C. Magn Reson Med Sci. 2012;11(4): 291–7.

27. Binkovitz LA, El-Youssef M, Glaser KJ, Yin M, Binkovitz AK, Ehman RL. Pediatric MR elastography of hepatic fibrosis: principles, technique and early clinical experience. Pediatr Radiol. 2012; 42(4):402–9.

28. Venkatesh SK, Teo LLS, Ang BWL, Lim SG, Wee A, Ehman RL. Effect of intravenous gadolinium on estimation of liver stiffness with MR elastography. European Congress of Radiology 2010. Vienna: European Society of Radiology; 2010. p. 163.

29. Motosugi U, Ichikawa T, Sou H, Sano K, Muhi A, Ehman RL, et al. Effects of gadoxetic acid on liver elasticity measurement by using magnetic resonance elastography. Magn Reson Imaging. 2012;30(1): 128–32.

30. Shire NJ, Yin M, Chen J, Railkar RA, Fox-Bosetti S, Johnson SM, et al. Test-retest repeatability of MR elastography for noninvasive liver fibrosis assessment in hepatitis C. J Magn Reson Imaging. 2011;34(4):947–55.

31. Hines CD, Bley TA, Lindstrom MJ, Reeder SB. Repeatability of magnetic resonance elastogra-

phy for quantification of hepatic stiffness. J Magn Reson Imaging. 2010;31(3):725–31.

32. Venkatesh SK, Wang G, Lim SG, Wee A. Magnetic resonance elastography for the detection and staging of liver fibrosis in chronic hepatitis B. Eur Radiol. 2014;24(1):70–8.

33. Godfrey EM, Patterson AJ, Priest AN, Davies SE, Joubert I, Krishnan AS, et al. A comparison of MR elastography and 31P MR spectroscopy with histological staging of liver fibrosis. Eur Radiol. 2012;22(12):2790–7.

34. Motosugi U, Ichikawa T, Sano K, Sou H, Muhi A, Koshiishi T, et al. Magnetic resonance elastography of the liver: preliminary results and estimation of inter-rater reliability. Jpn J Radiol. 2010;28(8):623–7.

35. Runge JH, Bohte AE, Verheij J, Terpstra V, Nederveen AJ, van Nieuwkerk KM, et al. Comparison of interobserver agreement of magnetic resonance elastography with histopathological staging of liver fibrosis. Abdom Imaging. 2014;39(2):283–90.

36. Blouin A, Bolender RP, Weibel ER. Distribution of organelles and membranes between hepatocytes and nonhepatocytes in the rat liver parenchyma. A stereological study. J Cell Biol. 1977;72(2):441–55.

37. Wisse E, Braet F, Luo D, De Zanger R, Jans D, Crabbé E, et al. Structure and function of sinusoidal lining cells in the liver. Toxicol Pathol. 1996;24(1):100–11.

38. Schuppan D. Structure of the extracellular matrix in normal and fibrotic liver: collagens and glycoproteins. Semin Liver Dis. 1990;10(1):1–10.

39. Rojkind M, Ponce-Noyola P. The extracellular matrix of the liver. Coll Relat Res. 1982;2(2):151–75.

40. Biagini G, Ballardini G. Liver fibrosis and extracellular matrix. J Hepatol. 1989;8(1):115–24.

41. Greenway C, Lautt WW. The hepatic circulation. In: Schultz S, editor. Handbook of physiology, vol. 1. Bethesda: American Physiological Society; 1989. p. 1519–64.

42. Nakata K. Microcirculation and hemodynamic analysis of the blood circulation of the liver. Acta Pathol Jpn. 1967;17(3):361–76.

43. Nakata K, Leong GF, Brauer RW. Direct measurement of blood pressures in minute vessels of the liver. Am J Physiol. 1960;199:1181–8.

44. Barrett KE, Boitano S, Barman SM, Brooks HL. Barrett K.E., Boitano S, Barman S.M., Brooks H.L. Chapter 28. Transport & Metabolic Functions of the Liver. In: Barrett KE, Boitano S, Barman SM, Brooks HL. Barrett K.E., Boitano S, Barman S.M., Brooks H.L. eds. Ganong's Review of Medical Physiology, 24e. New York, NY: McGraw-Hill; 2012. http://accessmedicine.mhmedical.com/content.aspx?bookid=393&Sectionid=39736772. Accessed August 07, 2014.

45. Rappaport AM. Microcirculatory units in the mammalian liver. Their arterial and portal components. Bibl Anat. 1977;16 Pt 2:116–20.

46. Lautt WW, Greenway CV. Hepatic venous compliance and role of liver as a blood reservoir. Am J Physiol. 1976;231(2):292–5.

47. Lautt WW, Greenway CV, Legare DJ. Index of contractility: quantitative analysis of hepatic venous distensibility. Am J Physiol. 1991;260(2 Pt 1):G325–32.

48. Lautt WW, Legare DJ. Passive autoregulation of portal venous pressure: distensible hepatic resistance. Am J Physiol. 1992;263(5 Pt 1):G702–8.

49. Bennett TD, Rothe CF. Hepatic capacitance responses to changes in flow and hepatic venous pressure in dogs. Am J Physiol. 1981;240(1):H18–28.

50. Venkatesh SK, Wang G, Teo LL, Ang BW. Magnetic resonance elastography of liver in healthy Asians: Normal liver stiffness quantification and reproducibility assessment. J Magn Reson Imaging. 2014;39(1):1–8.

51. Lee DH, Lee JM, Han JK, Choi BI. MR elastography of healthy liver parenchyma: Normal value and reliability of the liver stiffness value measurement. J Magn Reson Imaging. 2013;38(5):1215–23.

52. Kim BH, Lee JM, Lee YJ, Lee KB, Suh KS, Han JK, et al. MR elastography for noninvasive assessment of hepatic fibrosis: experience from a tertiary center in Asia. J Magn Reson Imaging. 2011;34(5):1110–6.

53. Herzka D, Kotys M, Sinkus R, Pettigrew R, Gharib A. Magnetic resonance elastography in the liver at 3 Tesla using a second harmonic approach. Magn Reson Med. 2009;62(2):284–91.

54. Mannelli L, Godfrey E, Graves MJ, Patterson AJ, Beddy P, Bowden D, et al. Magnetic resonance elastography: feasibility of liver stiffness measurements in healthy volunteers at 3T. Clin Radiol. 2012;67(3):258–62.

55. Hines CD, Lindstrom MJ, Varma AK, Reeder SB. Effects of postprandial state and mesenteric blood flow on the repeatability of MR elastography in asymptomatic subjects. J Magn Reson Imaging. 2011;33(1):239–44.

56. Pinzani M, Rombouts K. Liver fibrosis: from the bench to clinical targets. Dig Liver Dis. 2004;36(4):231–42.

57. Pinzani M, Rombouts K, Colagrande S. Fibrosis in chronic liver diseases: diagnosis and management. J Hepatol. 2005;42(Suppl(1)):S22–36.

58. Friedman SL. Liver fibrosis – from bench to bedside. J Hepatol. 2003;38 Suppl 1:S38–53.

59. Goodman Z, Becker RJ, Pockros P, Afdhal N. Progression of fibrosis in advanced chronic hepatitis C: evaluation by morphometric image analysis. Hepatology. 2007;45(4):886–94.

60. Shiffman ML, Stravitz RT, Contos MJ, Mills AS, Sterling RK, Luketic VA, et al. Histologic recurrence of chronic hepatitis C virus in patients after living

donor and deceased donor liver transplantation. Liver Transpl. 2004;10(10):1248–55.

61. Benhamou Y, Bochet M, Di Martino V, Charlotte F, Azria F, Coutellier A, et al. Liver fibrosis progression in human immunodeficiency virus and hepatitis C virus coinfected patients. The Multivirc Group. Hepatology. 1999;30(4):1054–8.

62. Batts KP, Ludwig J. Chronic hepatitis. An update on terminology and reporting. Am J Surg Pathol. 1995;19(12):1409–17.

63. Ishak K, Baptista A, Bianchi L, Callea F, De Groote J, Gudat F, et al. Histological grading and staging of chronic hepatitis. J Hepatol. 1995;22(6):696–9.

64. Bedossa P, Poynard T. An algorithm for the grading of activity in chronic hepatitis C. The METAVIR Cooperative Study Group. Hepatology. 1996; 24(2):289–93.

65. Brunt EM, Janney CG, Di Bisceglie AM, Neuschwander-Tetri BA, Bacon BR. Nonalcoholic steatohepatitis: a proposal for grading and staging the histological lesions. Am J Gastroenterol. 1999;94(9):2467–74.

66. Guido M. Chronic hepatitis: grading and staging. In: Saxena R, editor. Practical hepatic pathology: a diagnostic approach. St. Louis: WB Sanders; 2011. p. 201–13.

67. Standish RA, Cholongitas E, Dhillon A, Burroughs AK, Dhillon AP. An appraisal of the histopathological assessment of liver fibrosis. Gut. 2006;55(4): 569–78.

68. Venkatesh SK, Xu S, Tai D, Yu H, Wee A. Correlation of MR elastography with morphometric quantification of liver fibrosis (Fibro-C-Index) in chronic hepatitis B. Magn Reson Med. 2013. doi:10.1002/ mrm.25002

69. Vizzotto L, Vertemati M, Gambacorta M, Sabatella G, Spina V, Minola E. Analysis of histological and immunohistochemical patterns of the liver in posthepatitic and alcoholic cirrhosis by computerized morphometry. Mod Pathol. 2002;15(8):798–806.

70. Kage M, Shimamatu K, Nakashima E, Kojiro M, Inoue O, Yano M. Long-term evolution of fibrosis from chronic hepatitis to cirrhosis in patients with hepatitis C: morphometric analysis of repeated biopsies. Hepatology. 1997;25(4):1028–31.

71. Marcellin P, Ziol M, Bedossa P, Douvin C, Poupon R, de Lédinghen V, et al. Non-invasive assessment of liver fibrosis by stiffness measurement in patients with chronic hepatitis B. Liver Int. 2009;29(2):242–7.

72. Yano M, Kumada H, Kage M, Ikeda K, Shimamatsu K, Inoue O, et al. The long-term pathological evolution of chronic hepatitis C. Hepatology. 1996; 23(6):1334–40.

73. Silva AC, Walker W, Vargas V, Jatoi J, Ehman E, editors. Diffuse liver disease: virtual palpation with MR elastography. In: Proceedings of the radiological society of North America 93rd scientific assembly and annual meeting; 2008; Chicago.

74. Venkatesh SK, Yin M, Ehman RL. Magnetic resonance elastography of liver: clinical applications. J Comput Assist Tomogr. 2013;37(6):887–96.

75. Asbach P, Klatt D, Schlosser B, Biermer M, Muche M, Rieger A, et al. Viscoelasticity-based staging of hepatic fibrosis with multifrequency MR elastography. Radiology. 2010;257(1):80–6.

76. Chen J, Talwalkar JA, Yin M, Glaser KJ, Sanderson SO, Ehman RL. Early detection of nonalcoholic steatohepatitis in patients with nonalcoholic fatty liver disease by using MR elastography. Radiology. 2011;259(3):749–56.

77. Venkatesh SK, Yin M, Ehman RL. Magnetic resonance elastography of liver: technique, analysis, and clinical applications. J Magn Reson Imaging. 2013;37(3):544–55.

78. Bohte AE, van Dussen L, Akkerman EM, Nederveen AJ, Sinkus R, Jansen PLM, et al. Liver fibrosis in type I Gaucher disease: magnetic resonance imaging, transient elastography and parameters of iron storage. PLoS One. 2013;8(3):e57507.

79. Bensamoun SF, Leclerc GE, Debernard L, Cheng X, Robert L, Charleux F, et al. Cutoff values for alcoholic liver fibrosis using magnetic resonance elastography technique. Alcohol Clin Exp Res. 2013;37(5):811–7.

80. Crespo S, Bridges M, Nakhleh R, McPhail A, Pungpapong S, Keaveny AP. Non-invasive assessment of liver fibrosis using magnetic resonance elastography in liver transplant recipients with hepatitis C. Clin Transplant. 2013;27(5):652–8.

81. Myers RP, Crotty P, Pomier-Layrargues G, Ma M, Urbanski SJ, Elkashab M. Prevalence, risk factors and causes of discordance in fibrosis staging by transient elastography and liver biopsy. Liver Int. 2010;30(10):1471–80.

82. Singh S, Fujii LL, Murad MH, Wang Z, Asrani SK, Ehman RL, et al. Liver stiffness is associated with risk of decompensation, liver cancer, and death in patients with chronic liver diseases: a systematic review and meta-analysis. Clin Gastroenterol Hepatol. 2013;11(12):1573–84.e1-2. quiz e88-9.

83. Asrani SK, Talwalkar JA, Kamath PS, Shah VH, Saracino G, Jennings L, et al. Role of magnetic resonance elastography in compensated and decompensated liver disease. J Hepatol. 2013;60(5):934–9.

84. Sanyal AJ. NASH: a global health problem. Hepatol Res. 2011;41(7):670–4.

85. Younossi ZM, Stepanova M, Rafiq N, Makhlouf H, Younoszai Z, Agrawal R, et al. Pathologic criteria for nonalcoholic steatohepatitis: interprotocol agreement and ability to predict liver-related mortality. Hepatology. 2011;53(6):1874–82.

86. Rafiq N, Bai C, Fang Y, Srishord M, McCullough A, Gramlich T, et al. Long-term follow-up of patients with nonalcoholic fatty liver. Clin Gastroenterol Hepatol. 2009;7(2):234–8.

87. Musso G, Gambino R, Cassader M, Pagano G. Meta-analysis: natural history of non-alcoholic fatty liver disease (NAFLD) and diagnostic accuracy of non-invasive tests for liver disease severity. Ann Med. 2011;43(8):617–49.

88. Vernon G, Baranova A, Younossi ZM. Systematic review: the epidemiology and natural history of

non-alcoholic fatty liver disease and non-alcoholic steatohepatitis in adults. Aliment Pharmacol Ther. 2011;34(3):274–85.

89. Venkatesh SK, Wang G, Teo LL, Ang BW. Magnetic resonance elastography of liver in healthy asians: normal liver stiffness quantification and reproducibility assessment. J Magn Reson Imaging. 2014; 39(1):1–8.

90. Kim D, Kim WR, Talwalkar JA, Kim HJ, Ehman RL. Advanced fibrosis in nonalcoholic fatty liver disease: noninvasive assessment with MR elastography. Radiology. 2013;268(2):411–9.

91. Manabe N, Chevallier M, Chossegros P, Causse X, Guerret S, Trépo C, et al. Interferon-alpha 2b therapy reduces liver fibrosis in chronic non-A, non-B hepatitis: a quantitative histological evaluation. Hepatology. 1993;18(6):1344–9.

92. Poynard T, Mathurin P, Lai CL, Guyader D, Poupon R, Tainturier MH, et al. A comparison of fibrosis progression in chronic liver diseases. J Hepatol. 2003;38(3):257–65.

93. Caballero T, Pérez-Milena A, Masseroli M, O'Valle F, Salmerón FJ, Del Moral RM, et al. Liver fibrosis assessment with semiquantitative indexes and image analysis quantification in sustained-responder and non-responder interferon-treated patients with chronic hepatitis C. J Hepatol. 2001;34(5):740–7.

94. Sakaida I, Nagatomi A, Hironaka K, Uchida K, Okita K. Quantitative analysis of liver fibrosis and stellate cell changes in patients with chronic hepatitis C after interferon therapy. Am J Gastroenterol. 1999;94(2):489–96.

95. Duchatelle V, Marcellin P, Giostra E, Bregeaud L, Pouteau M, Boyer N, et al. Changes in liver fibrosis at the end of alpha interferon therapy and 6 to 18 months later in patients with chronic hepatitis C: quantitative assessment by a morphometric method. J Hepatol. 1998;29(1):20–8.

96. Friedman S, Bansal M. Reversal of hepatic fibrosis – fact or fantasy? Hepatology. 2006;43(2 Suppl 1):S82–8.

97. Dixon JB, Bhathal PS, Hughes NR, O'Brien PE. Nonalcoholic fatty liver disease: Improvement in liver histological analysis with weight loss. Hepatology. 2004;39(6):1647–54.

98. Chang TT, Liaw YF, Wu SS, Schiff E, Han KH, Lai CL, et al. Long-term entecavir therapy results in the reversal of fibrosis/cirrhosis and continued histological improvement in patients with chronic hepatitis B. Hepatology. 2010;52(3):886–93.

99. George SL, Bacon BR, Brunt EM, Mihindukulasuriya KL, Hoffmann J, Di Bisceglie AM. Clinical, virologic, histologic, and biochemical outcomes after successful HCV therapy: a 5-year follow-up of 150 patients. Hepatology. 2009;49(3):729–38.

100. Hoganson DD, Chen J, Ehman RL, Talwalkar JA, Michet CJJ, Yin M, et al. Magnetic resonance elastography for liver fibrosis in methotrexate treatment. Open J Rheumatol Autoimmune Dis. 2012;2:6–13.

101. Serai SD, Wallihan DB, Venkatesh SK, Ehman RL, Campbell KM, Sticka J, et al. Magnetic resonance elastography of the liver in patients status-post fontan procedure: feasibility and preliminary results. Congenit Heart Dis. 2014;9(1):7–14.

102. Lee VS, Miller FH, Omary RA, Wang Y, Ganger DR, Wang E, et al. Magnetic resonance elastography and biomarkers to assess fibrosis from recurrent hepatitis C in liver transplant recipients. Transplantation. 2011;92(5):581–6.

103. Kamphues C, Klatt D, Bova R, Yahyazadeh A, Bahra M, Braun J, et al. Viscoelasticity-based magnetic resonance elastography for the assessment of liver fibrosis in hepatitis C patients after liver transplantation. Rofo. 2012;184(11):1013–9.

104. Mariappan YK, Venkatesh SK, Glaser JK, McGee KP, Ehman RL, editors. MR elastography of liver with iron overload: development, evaluation and preliminary clinical experience with improved spin echo and spin echo EPI sequences. In: The 21st annual conference of international society for magnetic resonance in medicine; 2013; Salt Lake City, Utah.

105. Motosugi U, Ichikawa T, Amemiya F, Sou H, Sano K, Muhi A, et al. Cross-validation of MR elastography and ultrasound transient elastography in liver stiffness measurement: discrepancy in the results of cirrhotic liver. J Magn Reson Imaging. 2012; 35(3):607–10.

106. Oudry J, Chen J, Glaser K, Miette V, Sandrin L, Ehman R. Cross-validation of magnetic resonance elastography and ultrasound-based transient elastography: a preliminary phantom study. J Magn Reson Imaging. 2009;30(5):1145–50.

107. Yoon JH, Lee JM, Woo HS, Yu MH, Joo I, Lee ES, et al. Staging of hepatic fibrosis: comparison of magnetic resonance elastography and shear wave elastography in the same individuals. Korean J Radiol. 2013;14(2):202–12.

108. Guo Y, Parthasarathy S, Goyal P, McCarthy RJ, Larson AC, Miller FH. Magnetic resonance elastography and acoustic radiation force impulse for staging hepatic fibrosis: a meta-analysis. Abdom Imaging. 2014. (doi:10.1007/s00261-014-0137-6).

109. Venkatesh SK, Teo LLS, Ang BWL, Ehman RL, editors. Non-invasive detection of liver fibrosis: comparison of MR elastography with diffusion weighted MR imaging. In: Proceedings of ECR 2010; Vienna: European Society of Radiology; 2010.

110. Wang Y, Ganger DR, Levitsky J, Sternick LA, McCarthy RJ, Chen ZE, et al. Assessment of chronic hepatitis and fibrosis: comparison of MR elastography and diffusion-weighted imaging. AJR Am J Roentgenol. 2011;196(3):553–61.

111. Wang QB, Zhu H, Liu HL, Zhang B. Performance of magnetic resonance elastography and diffusion-weighted imaging for the staging of hepatic fibrosis: a meta-analysis. Hepatology. 2012;56(1):239–47.

112. Choi YR, Lee JM, Yoon JH, Han JK, Choi BI. Comparison of magnetic resonance elastography

and gadoxetate disodium-enhanced magnetic resonance imaging for the evaluation of hepatic fibrosis. Invest Radiol. 2013;48(8):607–13.

113. Motosugi U, Ichikawa T, Muhi A, Sano K, Morisaka H, Ichikawa S, et al. Magnetic resonance elastography as a predictor of insufficient liver enhancement on gadoxetic acid-enhanced hepatocyte-phase magnetic resonance imaging in patients with type C hepatitis and Child-Pugh class A disease. Invest Radiol. 2012;47(10):566–70.

114. Noren B, Dahlqvist O, Lundberg P, Almer S, Kechagias S, Ekstedt M, et al. Separation of advanced from mild fibrosis in diffuse liver disease using 31P magnetic resonance spectroscopy. Eur J Radiol. 2008;66(2):313–20.

115. Corbin IR, Ryner LN, Singh H, Minuk GY. Quantitative hepatic phosphorus-31 magnetic resonance spectroscopy in compensated and decompensated cirrhosis. Am J Physiol Gastrointest Liver Physiol. 2004;287(2):G379–84.

116. Dezortova M, Taimr P, Skoch A, Spicak J, Hajek M. Etiology and functional status of liver cirrhosis by 31P MR spectroscopy. World J Gastroenterol. 2005;11(44):6926–31.

Clinical Applications of Liver Magnetic Resonance Elastography: Focal Liver Lesions

5

Sudhakar K. Venkatesh

Introduction

Focal liver lesions (FLLs) are detected at an increasing rate due to widespread use of imaging for abdominal pain and other symptoms. Accurate and reliable characterization of the FLL is critical for both reassuring individuals with benign lesions and accurately diagnosing malignant lesions for management [1]. Correct diagnosis would avoid delayed treatment of a malignancy and probably reduce the unnecessary treatment of benign lesions.

Accurate characterization of FLL with conventional imaging methods is challenging due to frequent overlap of imaging features. As a result, some FLLs that are difficult to characterize and in those where no diagnosis can be reached on the basis of imaging alone, histological confirmation may be required for further management. However, biopsy is invasive with a non-negligible risk of seeding when a malignant lesion is biopsied. Distinction between benign and malignant is perhaps the most important information needed for clinical decisions, particularly when FLLs are asymptomatic or an incidental finding. The current imaging approach for characterization of FLL is qualitative and evaluates the differences between FLL and the surrounding liver parenchyma, and observance of contrast enhancement characteristics on dynamic multiphasic images. Dynamic contrast enhanced imaging with CT and MRI puts patients at risk of exposure to ionizing radiation with CT and the possible side effects of contrast medium administration especially in patients with chronic renal disease.

Diffusion weighted imaging (DWI) is a sensitive imaging method and useful in differentiating benign and malignant FLLs and shown to be better than standard T2-weighted imaging [2] but similar to dynamic contrast enhanced MRI [3]. However apparent diffusion coefficient (ADC) is not specific enough to be used clinically. Intravoxel incoherent motion (IVIM) imaging derived parameters such as true diffusion may be a better parameter for differentiating benign and malignant FLLs as perfusion influences ADC values [4].

Newer contrast media such as Gadoxetate sodium are used for differentiation of hepatocellular lesions from non-hepatocellular lesions. However use of a contrast agent in patients with severe renal failure may not be always possible even though 50 % of Gadoxetate is excreted through liver. Gadoxetate is no longer considered discriminative as once thought to be, as some well differentiated hepatocellular carcinomas (HCC) and hepatic adenomas show uptake. MRI with super paramagnetic iron oxide agents are very sensitive to detect FLLs, however the agent is no longer available for clinical use in USA.

S.K. Venkatesh (✉)
Department of Radiology, Mayo Clinic College
of Medicine, Mayo clinic, 200 First Street SW,
Rochester, MN, USA
e-mail: venkatesh.sudhakar@mayo.edu

S.K. Venkatesh and R.L. Ehman (eds.), *Magnetic Resonance Elastography*,
DOI 10.1007/978-1-4939-1575-0_5, © Springer Science+Business Media New York 2014

Tumors can be palpated as hard or stiff masses within softer normal tissues. However tumors within deep seated organs like liver cannot be assessed with palpation. Elastography techniques may be useful in assessing stiffness of the tumors and in differentiating them from normal tissues. Elastography techniques by quantifying stiffness can provide for differentiation of benign and malignant FLLs [5–7]. Transient elastography (TE) is not an appropriate technique as it does not provide any imaging guidance or precise location of sampling of the stiffness. Acoustic radiation force impulse (ARFI) imaging combines ultrasound imaging and elastography and able to evaluate focal lesions. However ARFI provides an average stiffness value over a fixed region of interest which is usually not deeper than 6 cm from skin surface [8], may produce inaccurate results when FLLs are near vessels or heart [9], and shear wave velocities for the same kind of lesion may differ depending on the surrounding liver parenchyma [10, 11]. Shear wave elastography (SWE) is another ultrasound based technique reported to be useful for evaluation of FLL [12] but is limited by significant failure rate and depth limitation. Experience with this technique is still growing.

MRE is probably the most suitable technique for evaluation of stiffness of FLLs as it can provide exact localization of the tumor and is not limited by size or depth of the lesion. Two studies have reported evaluation of FLL with MRE [5, 6] and both studies have reported moderate to excellent accuracy for differentiation of benign and malignant FLLs. The preliminary results are encouraging and provide motivation for further evaluation of MRE in this field.

Tumor Stiffness

Cancers are stiffer than normal tissue. Several factors including tissue composition, abnormal perfusion, and higher interstitial pressure (IP) may explain the stiffness of cancers. Cancer cells often occupy less than half the volume of a solid tumor, blood vessels account for about 10 % of volume, and the remaining space is made of interstitial stroma that surrounds the cancers cells [13]. The tumor interstitial stroma is characterized by abnormal extracellular matrix rich in collagen, increased micro vessel density, and activated fibroblasts and an inflammatory cell infiltrate [14]. The tumor vasculature is heterogeneous and the vessels are irregular, convoluted with increased vascular permeability leading to abnormal flow and leaky vessels with increased outflow of proteins and other molecules out of the vessels and increasing the osmotic pressure of the interstitium [15–19]. The lymphatic vessels are either non-functional or absent [20, 21]; as a result there is inefficient drainage of fluid from tumor. All these contribute to increased IP accompanied by increased micro vascular pressure [22]. Tumor IP can be several times that of normal tissues [23]. For example, colorectal liver metastases and normal liver have a mean interstitial pressure of 21 and 7–10 mmHg respectively, and breast carcinomas have mean interstitial pressure of 15 versus 0 mmHg in normal breast [13, 24]. The high IP is probably due to micro vascular pressure, blood-vessel leakiness, lymph-vessel abnormalities, interstitial fibrosis, and contraction of the interstitial space mediated by stromal fibroblasts [14, 25–27]. The tumor IP is dependent on host vasculature and positively correlates with blood volume of the tumor and negatively with apoptosis/necrosis [28–30]. The high IP is uniform throughout a solid tumor except at the rim and interface with normal tissue where it reaches normal tissue interstitial pressure [13]. As cancers grow, several nodules may form with their own smaller micro-tumor environments as well as areas of necrosis which may have lower IP. Larger tumors may therefore have variable IP in different regions [24, 31].

Tumor IP has been the focus of cancer treatment research as studies have shown that IP is the main barrier for delivery of anti-cancer agents and reducing the tumor IP helps in delivery of anti-cancer agents [13, 14, 17, 19, 22, 24, 29–31]. Further, a decrease in tumor IP is a known marker of response to therapy [30]. An expanding tumor also exerts stresses on the surrounding normal tissue. These stresses are transmitted to the surrounding extracellular matrix, generating radial compressive forces and circumferential tensile forces. The surrounding matrix becomes stiffer and denser, the degree of which is variable [32] making the tumor environment complex. In summary, the tumor stiffness is a result of many factors including high cellularity, collagen rich extracellular matrix, altered micro vessels, raised IP, growing stress on surrounding tissues, and reaction of surrounding tissues.

MRE of FLLs: Technique

FLLs are evaluated with the same technique as for evaluation of diffuse liver diseases described in previous chapters. To ensure valid evaluation of stiffness of FLL, the slices need to be planned so that MRE acquisition is through the plane of the tumor and preferably through the largest cross-section or closer to the center of the FLL. Thin (3–5 mm) slices may be obtained, but this may reduce the signal from the tissues and therefore MRE signal. The MRE acquisition may be performed after intravenous gadolinium administration which would increase the signal from tissues especially enhancing FLLs. Matrix size of the MRE acquisition may also be increased to obtain higher resolution. MRE of smaller lesions may be technically challenging especially those <2 cm as thinner slices and consistent breath hold are required to obtain slices through the small lesion. Smaller lesions are better evaluated with a 3D MRE technique to ensure reliable stiffness estimation. A higher frequency (90–200 Hz) (therefore smaller wavelength shear waves) may be useful for evaluating small lesions. However, higher frequency waves are attenuated in normal liver [5] but stiffer tissues may be still visualized.

MRE of FLLs

The stiffness of FLLs is dependent on tissue components. Benign tumors are softer and have slightly higher stiffness than normal liver parenchyma (Fig. 5.1). Some focal nodular hyperplasia may be stiffer probably due to their central scar and fibrous component (Fig. 5.1). Hemangiomas

are lesions composed of large vascular spaces and fibrous septa that are typically softer. Malignant tumors have higher stiffness (Fig. 5.2) and some tumors like cholangiocarcinoma demonstrate higher stiffness probably due to their desmoplastic reaction they induce and the increased amount of fibrotic tissues. The stiffness of metastatic lesions can be variable and dependent on their components. Most solid metastases are stiffer than normal liver.

In a preliminary study, Venkatesh et al. [5] studied 44 focal tumors (13 benign and 31 malignant) and found that the mean stiffness of the malignant lesions was significantly higher than benign tumors (10.1 vs. 2.7 kPa, $p < 0.001$), normal liver parenchyma (vs. 2.3 kPa, $p < 0.001$), and fibrotic livers (vs. 5.9 kPa, $p < 0.001$) (Fig. 5.3). The mean stiffness of benign tumors was not significantly different from normal liver parenchyma. Fibrotic livers were significantly stiffer than benign liver tumors. In this study, a threshold value of 5 kPa was able to accurately distinguish malignant from benign liver tumors (Fig. 5.3) with 100 % accuracy. Garteiser et al. [6] studied 72 FLLs (37 benign and 35 malignant) and found that absolute shear modulus and loss modulus were significantly higher in malignant tumors as compared to benign tumors. The loss modulus was found to be significantly better than storage modulus and shear modulus (0.774 vs 0.578 and vs.0.718 respectively) for discriminating malignant from benign lesions. These preliminary studies need to be supported by larger studies. In a recent study, Venkatesh et al. showed that MRE performs significantly better than DWI in differentiating primary neoplasms of the liver (0.98 vs.0.85, $p = 0.008$) [33].

Malignant tumors tend to be larger than benign tumors. However the tissue stiffness is dependent on composition and therefore stiffness of the FLL is probably independent of the size of the lesion (Fig. 5.4). Among the benign tumors, hemangiomas and hepatocellular adenomas are softer than focal nodular hyperplasia. Cholangiocarcinomas are stiffer than HCCs and may be dependent on relative amount of fibrosis, solid tissue, and necrotic components. Characterization of individual types of FLL using stiffness values may be possible and

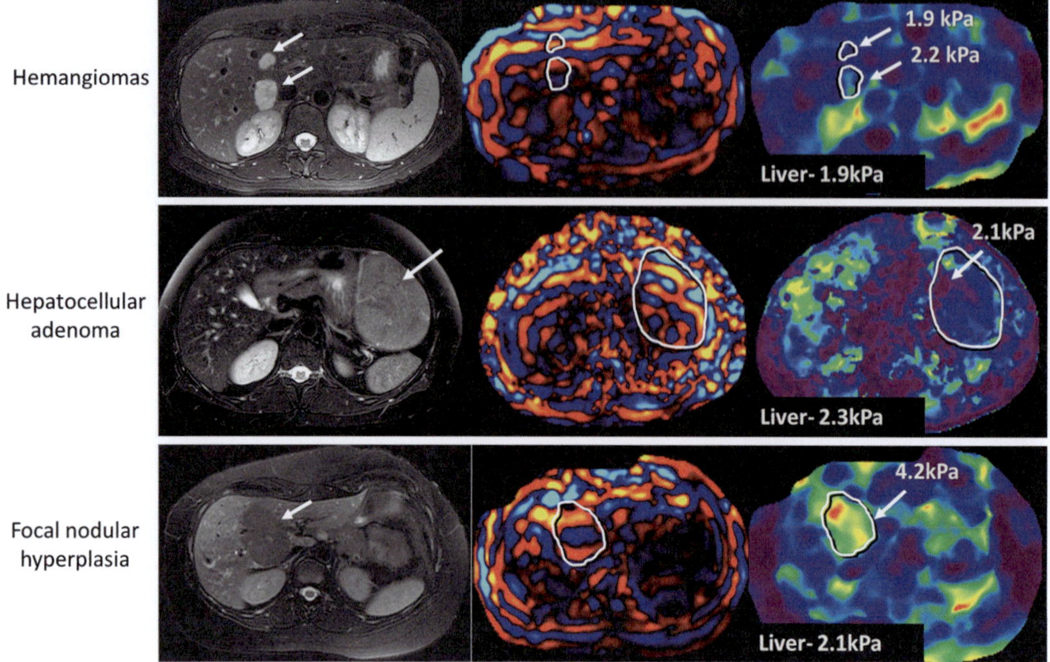

Fig. 5.1 MRE of benign FLLs. Examples of hemangioma (*top row*), hepatocellular adenoma (*middle row*), and focal nodular hyperplasia (*bottom row*). T2-weighted images (*first column*), wave images (*second column*) and stiffness maps (*third column*). The mean stiffness of hemangioma and hepatocellular adenoma is similar to slightly higher than liver parenchyma. Focal nodular hyperplasia demonstrated here has higher stiffness probably due to fibrotic scaring but is still less than 5 kPa

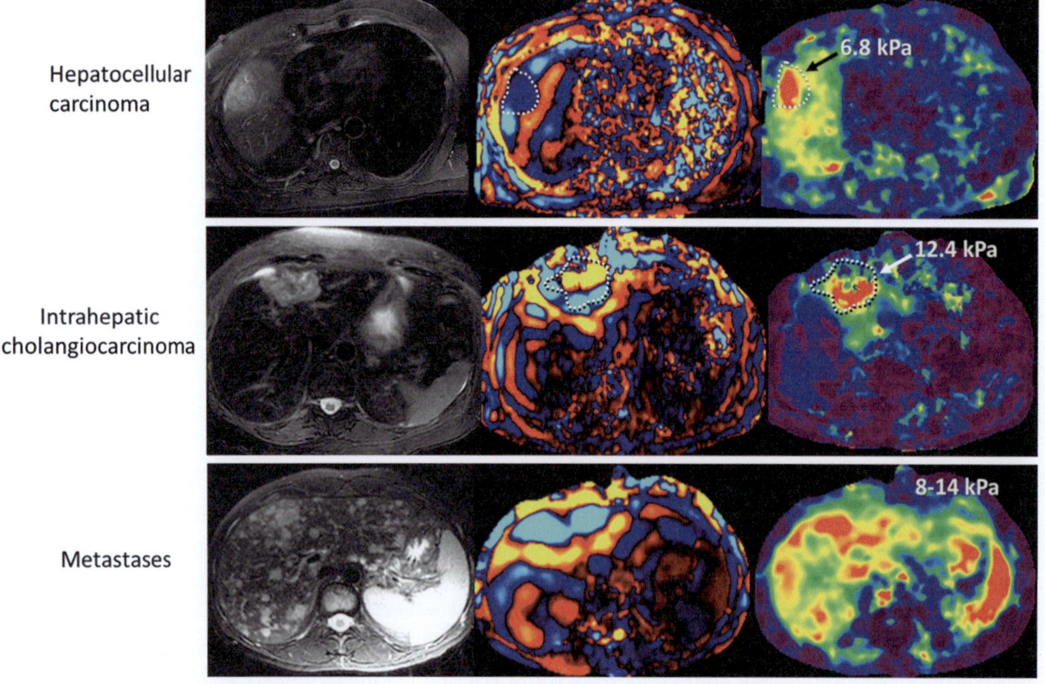

Fig. 5.2 MRE of malignant FLLs. T2-weighted images (*first column*), wave images (*second column*) and stiffness maps (*third column*). Examples of hepatocellular carcinoma, intrahepatic cholangiocarcinoma, and metastases. The mean stiffnesses of the lesions are significantly higher than the surrounding liver. Note the increased stiffness of the liver in multiple metastases with focal areas of increased stiffness representing the metastases

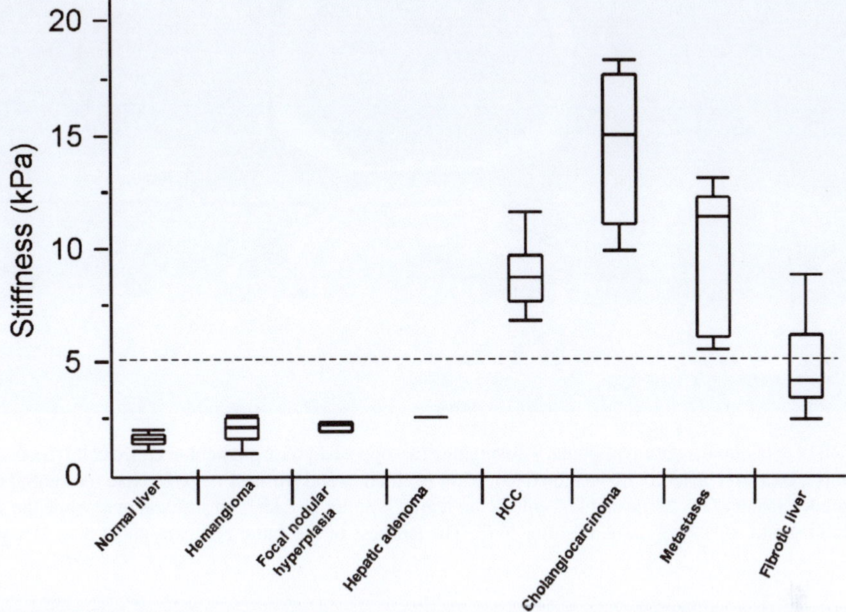

Fig. 5.3 *Box-plot graph* showing distribution of mean stiffness of 44 benign and malignant focal liver lesions, normal liver, and fibrotic liver. Note that a cut-off value of 5 kPa separates malignant and benign tumors. Fibrotic liver stiffness overlaps the benign and malignant tumors. (Reproduced with permission from Venkatesh et al. [5]: Reprinted with permission from the American Journal of Roentgenology)

Fig. 5.4 Graph showing correlation between size and stiffness of 110 focal liver lesions (malignant-85 and benign-25). *Red dots* represent malignant tumors and *green dots* benign tumors. No significant correlation was found between stiffness and tumor size ($R^2 = 0.05$). Three sclerosing hemangiomas (*arrows*) had stiffness more than 5 kPa. (Presented at SBCTMR Annual Conference 2008, Charleston, SC, USA) (color figure online)

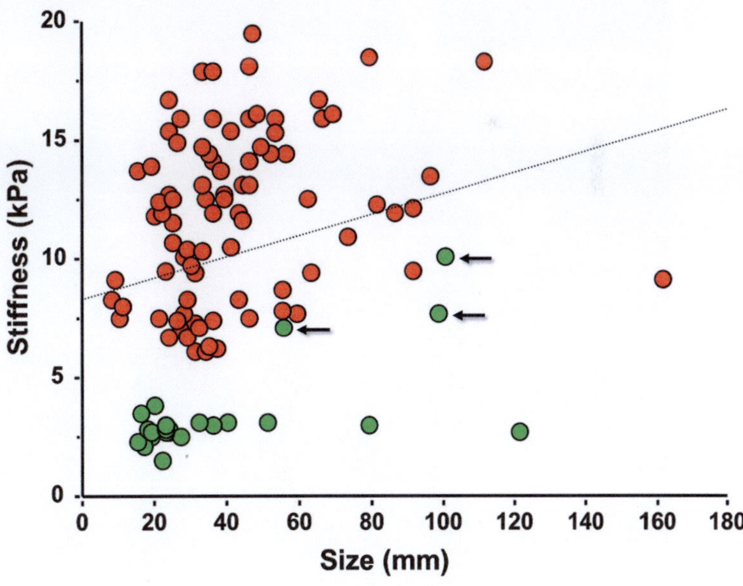

future studies in this direction are awaited. An example is demonstrated in Fig. 5.5. In this example of epithelioid angiomyolipoma, the lesion shows a predominantly fatty core surrounded by an intensely enhancing region. The central core region is softer while the surrounding region is stiffer.

Cystic or necrotic regions of large tumors, or those which become necrotic following treatment, may be demonstrated as areas of lower stiffness (Fig. 5.6). In a recent study, Pepin et al. demonstrated a significant decrease in tumor shear stiffness within 4 days of chemotherapy

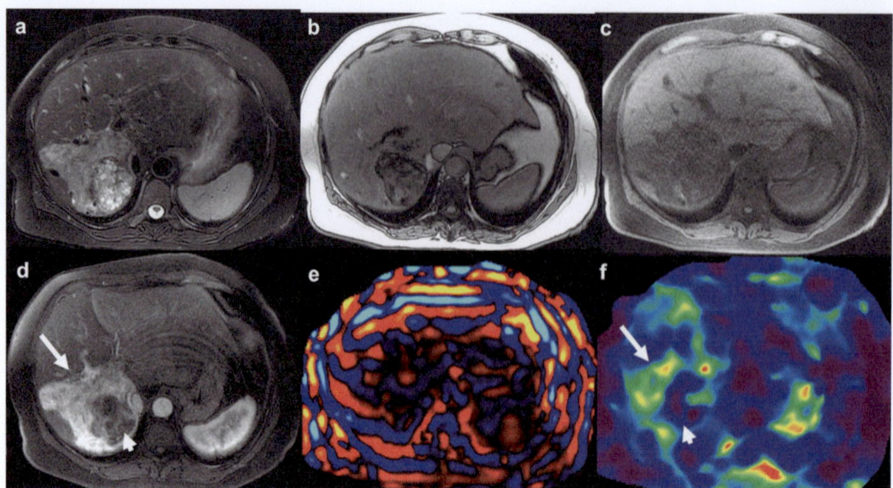

Fig. 5.5 MRE of epithelioid angiomyolipoma. T2-weighted (**a**), opposed-phase gradient–recalled echo (**b**), fat-suppressed T1-weighted (**c**), contrast enhanced T1-weighted (**d**), wave (**e**) images and stiffness map (**f**). Note the central core (*arrow head*) with predominantly fatty tissue and less enhancing region is softer (1.8 kPa) on stiffness map while the surrounding less fat containing and enhancing tissue is stiffer (4.6). The stiffness of the entire FLL was about 3.18 kPa suggesting a benign FLL

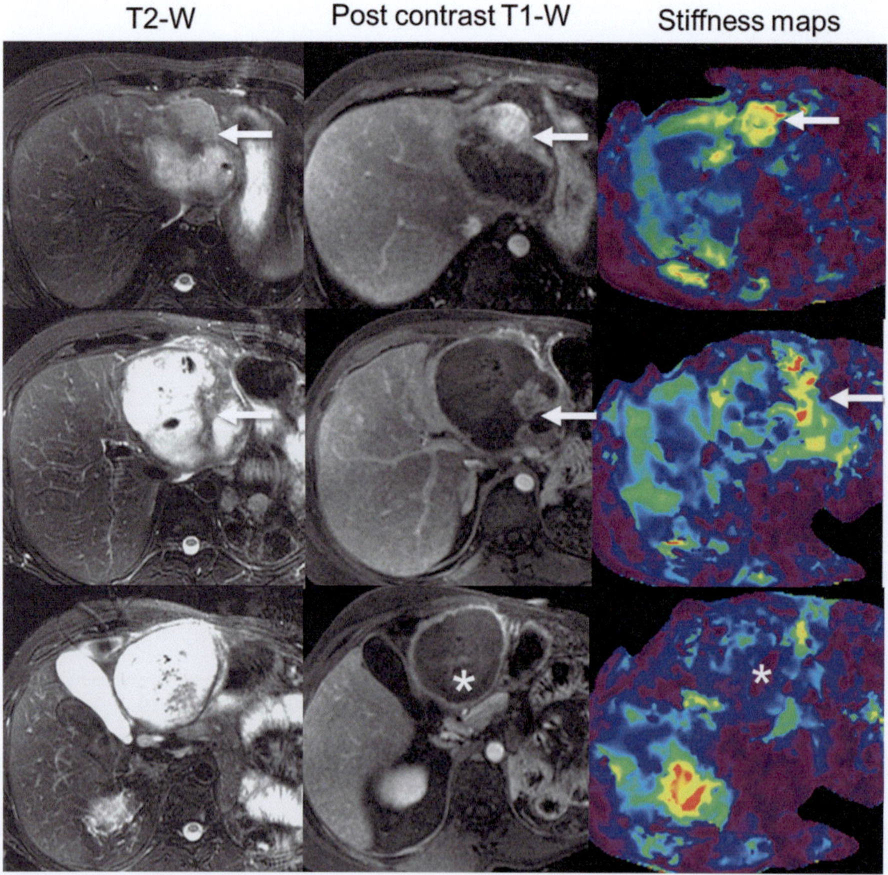

Fig. 5.6 MRE of hepatocellular carcinoma 6 weeks post chemoembolization. The large mass in left lobe liver shows significant necrosis with residual enhancing regions (*arrows*) consistent with residual tumor. The residual tumors are seen as stiff regions on the stiffness maps. Section through a completely necrotic region (*asterisk, bottom row*) shows low stiffness

treatment in an animal model. A similar study by Li et al. [34] also demonstrated that a reduction in tumor viscoelasticity occurs earlier than changes in tumor ADC. Demonstration of change in MRE-derived shear stiffness may be useful as an early and sensitive biomarker of tumor response. This early reduction in stiffness may represent reduced vascularity and decreased interstitial pressure that can occur before necrosis results.

development [35, 36]; however in another study Anaparthy et al. [37] found no such correlation. Clearly more studies are warranted in this direction to determine whether stiffness of liver is a risk factor for development of HCC. HCCs developing in cirrhotic liver may appear either stiffer or softer than surrounding parenchyma depending on the degree of fibrosis in the surrounding parenchyma and components of the HCC (Fig. 5.7).

Focal Lesions in Fibrotic Liver

Fibrosis of liver is a known risk factor for development of HCC. As stiffness correlates with degree of fibrosis, elevated stiffness may be considered a risk factor for development of HCC. Studies have shown that liver stiffness assessed with MRE or ultrasound based technique is associated with higher risk of HCC

Other Stiff Focal Lesions in Liver

Benign lesions that contain fibrotic tissue components may be stiffer and these should not be interpreted as malignant lesions. Lesions that may appear as stiff regions include inflammatory lesions which have fibrotic components such as confluent fibrosis, hyalinizing granulomas, and healing liver abscesses (Fig. 5.8). Some benign

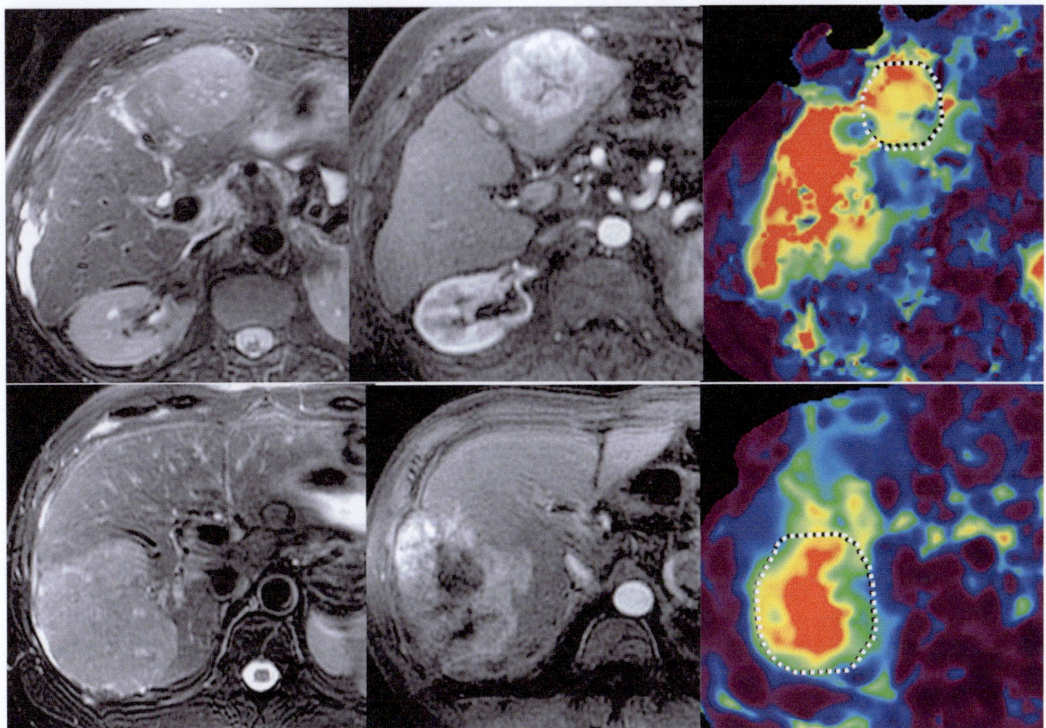

Fig. 5.7 MRE of hepatocellular carcinomas in fibrotic liver. T2-weighted images (*first column*), late arterial phase (*second column*) and stiffness maps (*third column*). In the first example (*top row*) the HCC has a stiffness of 6.6 kPa and surrounding liver parenchyma is 8.3 kPa. In the *bottom row*, the stiffness of HCC and liver parenchyma are 7.6 and 3.4 kPa respectively

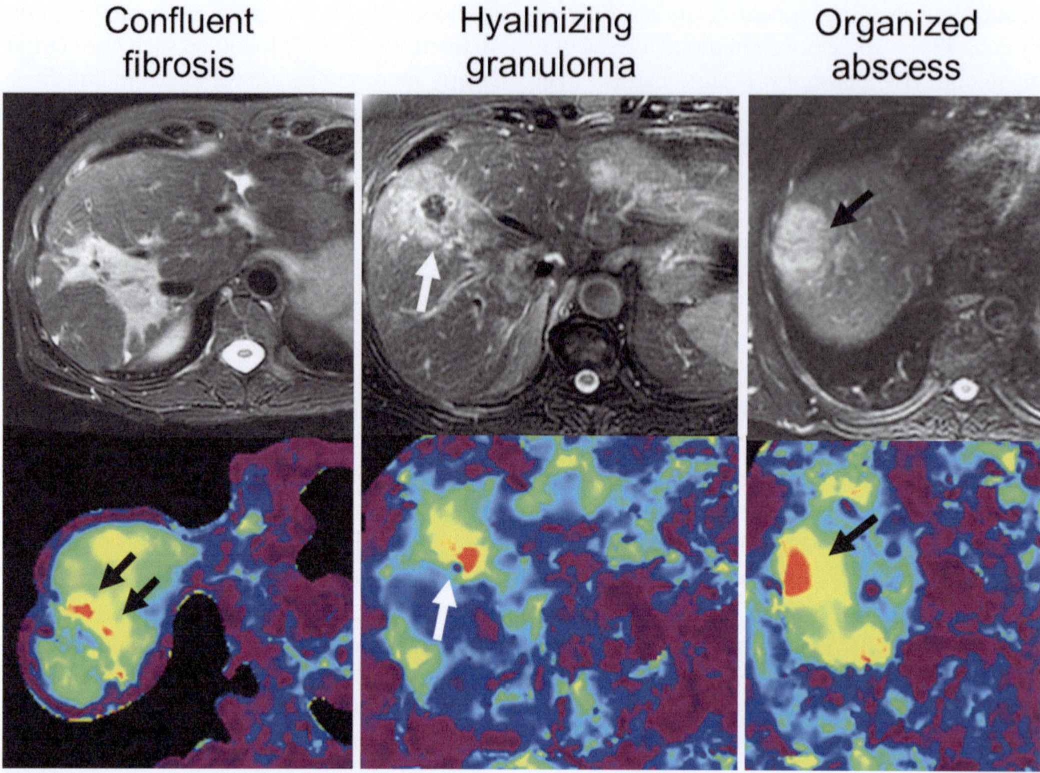

Fig. 5.8 Benign focal lesions that can present as stiff regions on MRE. T2-weighted images (*top row*) and stiffness maps (*bottom row*). Examples of confluent fibrosis, hyalinizing granuloma, and organized abscess demonstrating increased stiffness (*arrows*)

tumors like focal nodular hyperplasia (Fig. 5.1) and sclerosing hemangiomas can appear very stiff (Fig. 5.9) as they can have stiffer components. The stiffness values of these lesions should be interpreted with other imaging features for characterization. Shear waves do not propagate through cysts as fluids do not support shear waves. Therefore cysts typically appear as soft lesions or void regions within liver.

Future Directions

Evaluating stiffness of small FLLs with MRE can be challenging and 3D MRE may be useful to avoid partial volume effects and the edge effects

in the inversion algorithm. Using minimally attenuating longitudinal waves at a high frequency, shear waves produced by mode conversion will be detectable in stiff regions (Fig. 5.10) but attenuated in the surrounding softer tissue [38]. This interesting and novel technique may be useful to detect smaller lesions, however experience is limited.

MRE is a promising technique for evaluation of FLLs and provides a non-invasive quantitative parameter that may be useful in differentiating benign and malignant neoplasms. Change in the MRE derived stiffness may be useful in detection of early tumor response to treatment. 3D MRE techniques may allow evaluation of smaller lesions.

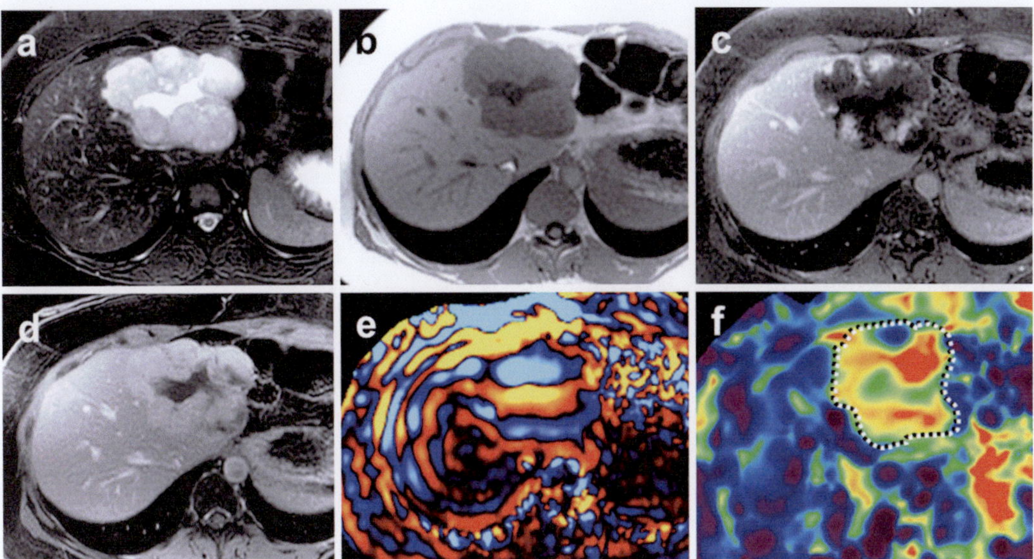

Fig. 5.9 MRE of a sclerosing hemangioma. T2-weighted (**a**), T1-weighted (**b**), post contrast enhanced (**c**) and delayed (**d**) T1-weighted images, wave image (**e**), and stiffness (**f**) showing a large hemangioma in the left lobe that shows increased stiffness with MRE

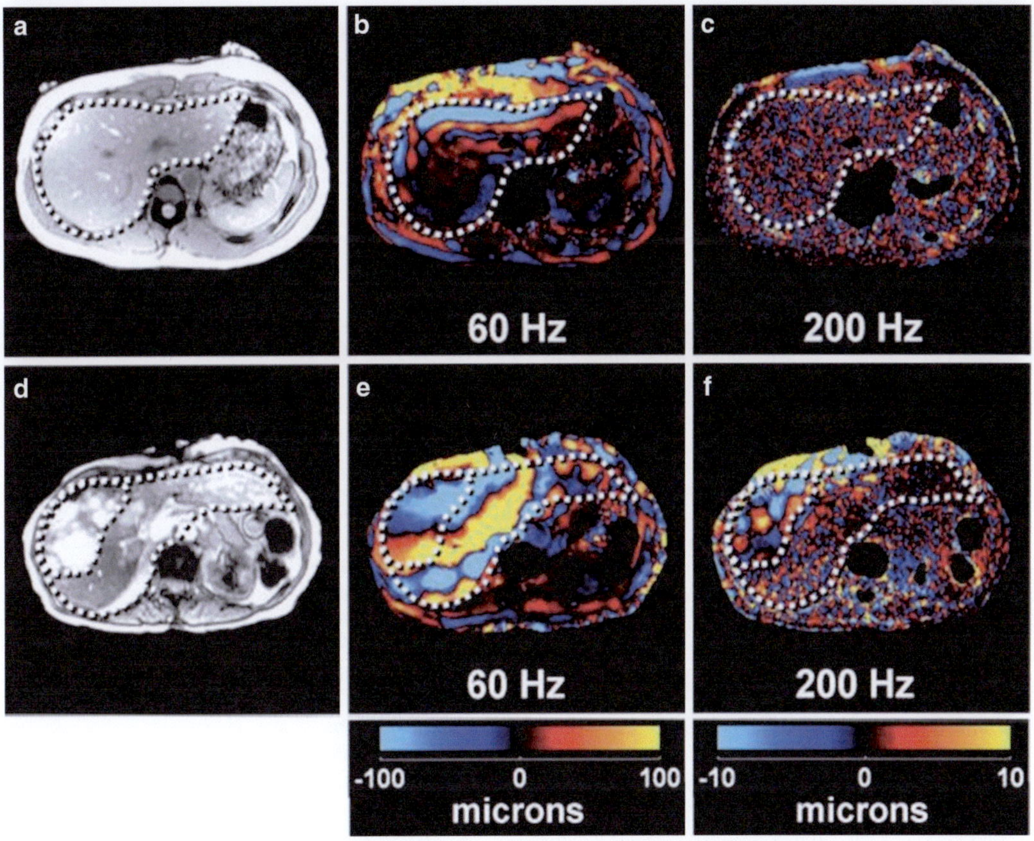

Fig. 5.10 Results from an in vivo hepatic study of a normal volunteer and a patient with metastatic liver tumors. Conventional MR magnitude images of the volunteer (**a**) and the patient (**d**) show the presence of numerous tumors in the patient. Wave data in the volunteer at 60 Hz (**b**) and 200 Hz (**c**) show that the shear waves at 200 Hz are completely attenuated in normal healthy soft liver tissue. The wave data in the patient at 60 Hz (**e**) and 200 Hz (**f**) show that the waves in the stiff tumors are still clearly observable while the shear waves in the rest of the liver have been significantly attenuated. (Reproduced with permission from Mariappan et al. [38])

References

1. Venkatesh SK, Chandan V, Roberts LR. Liver masses: a clinical, radiologic, and pathologic perspective. Clin Gastroenterol Hepatol. 2013. doi:10.1016/j.cgh.2013.09.017.
2. Parikh T, Drew SJ, Lee VS, Wong S, Hecht EM, Babb JS, et al. Focal liver lesion detection and characterization with diffusion-weighted MR imaging: comparison with standard breath-hold T2-weighted imaging. Radiology. 2008;246(3):812–22.
3. Tappouni R, ElKady RM, Sarwani N, Dykes T. Comparison of the accuracy of diffusion-weighted imaging versus dynamic contrast enhancement magnetic resonance imaging in characterizing focal liver lesions. J Comput Assist Tomogr. 2013;37(6): 995–1001.
4. Yoon JH, Lee JM, Yu MH, Kiefer B, Han JK, Choi BI. Evaluation of hepatic focal lesions using diffusion-weighted MR imaging: comparison of apparent diffusion coefficient and intravoxel incoherent motion-derived parameters. J Magn Reson Imaging. 2014;39(2):276–85.
5. Venkatesh SK, Yin M, Glockner JF, Takahashi N, Araoz PA, Talwalkar JA, et al. MR elastography of liver tumors: preliminary results. AJR Am J Roentgenol. 2008;190(6):1534–40.
6. Garteiser P, Doblas S, Daire JL, Wagner M, Leitao H, Vilgrain V, et al. MR elastography of liver tumours: value of viscoelastic properties for tumour characterisation. Eur Radiol. 2012;22(10):2169–77.
7. Ma X, Zhan W, Zhang B, Wei B, Wu X, Zhou M, et al. Elastography for the differentiation of benign and malignant liver lesions: a meta-analysis. Tumour Biol. 2014;35(5):4489–97.
8. Fahey BJ, Nightingale KR, Nelson RC, Palmeri ML, Trahey GE. Acoustic radiation force impulse imaging of the abdomen: demonstration of feasibility and utility. Ultrasound Med Biol. 2005;31(9):1185–98.
9. Cho SH, Lee JY, Han JK, Choi BI. Acoustic radiation force impulse elastography for the evaluation of focal solid hepatic lesions: preliminary findings. Ultrasound Med Biol. 2010;36(2):202–8.
10. Ying L, Lin X, Xie ZL, Tang FY, Hu YP, Shi KQ. Clinical utility of acoustic radiation force impulse imaging for identification of malignant liver lesions: a meta-analysis. Eur Radiol. 2012;22(12):2798–805.
11. Gallotti A, D'Onofrio M, Romanini L, Cantisani V, Pozzi MR. Acoustic radiation force impulse (ARFI) ultrasound imaging of solid focal liver lesions. Eur J Radiol. 2012;81(3):451–5.
12. Guibal A, Boularan C, Bruce M, Vallin M, Pilleul F, Walter T, et al. Evaluation of shearwave elastography for the characterisation of focal liver lesions on ultrasound. Eur Radiol. 2013;23(4):1138–49.
13. Jain RK. Barriers to drug delivery in solid tumors. Sci Am. 1994;271(1):58–65.
14. Heldin C-H, Rubin K, Pietras K, Östman A. High interstitial fluid pressure—an obstacle in cancer therapy. Nat Rev Cancer. 2004;4(10):806–13.
15. Baluk P, Morikawa S, Haskell A, Mancuso M, McDonald DM. Abnormalities of basement membrane on blood vessels and endothelial sprouts in tumors. Am J Pathol. 2003;163(5):1801–15.
16. Morikawa S, Baluk P, Kaidoh T, Haskell A, Jain RK, McDonald DM. Abnormalities in pericytes on blood vessels and endothelial sprouts in tumors. Am J Pathol. 2002;160(3):985–1000.
17. Jain RK. Delivery of molecular and cellular medicine to solid tumors. Adv Drug Deliv Rev. 2001;46(1–3): 149–68.
18. Dvorak HF, Brown LF, Detmar M, Dvorak AM. Vascular permeability factor/vascular endothelial growth factor, microvascular hyperpermeability, and angiogenesis. Am J Pathol. 1995;146(5):1029–39.
19. Jain RK. Normalizing tumor vasculature with anti-angiogenic therapy: a new paradigm for combination therapy. Nat Med. 2001;7(9):987–9.
20. Padera TP, Kadambi A, di Tomaso E, Carreira CM, Brown EB, Boucher Y, et al. Lymphatic metastasis in the absence of functional intratumor lymphatics. Science. 2002;296(5574):1883–6.
21. Alitalo K, Carmeliet P. Molecular mechanisms of lymphangiogenesis in health and disease. Cancer Cell. 2002;1(3):219–27.
22. Boucher Y, Jain RK. Microvascular pressure is the principal driving force for interstitial hypertension in solid tumors: implications for vascular collapse. Cancer Res. 1992;52(18):5110–4.
23. Young JS, Lumsden CE, Stalker AL. The significance of the tissue pressure of normal testicular and of neoplastic (Brown-Pearce carcinoma) tissue in the rabbit. J Pathol Bacteriol. 1950;62(3):313–33.
24. Less JR, Posner MC, Boucher Y, Borochovitz D, Wolmark N, Jain RK. Interstitial hypertension in human breast and colorectal tumors. Cancer Res. 1992;52(22):6371–4.
25. Sevick EM, Jain RK. Geometric resistance to blood flow in solid tumors perfused ex vivo: effects of tumor size and perfusion pressure. Cancer Res. 1989;49(13): 3506–12.
26. Sevick EM, Jain RK. Viscous resistance to blood flow in solid tumors: effect of hematocrit on intratumor blood viscosity. Cancer Res. 1989;49(13):3513–9.
27. Less JR, Skalak TC, Sevick EM, Jain RK. Microvascular architecture in a mammary carcinoma: branching patterns and vessel dimensions. Cancer Res. 1991;51(1):265–73.
28. Lunt SJ, Kalliomaki TM, Brown A, Yang VX, Milosevic M, Hill RP. Interstitial fluid pressure, vascularity and metastasis in ectopic, orthotopic and spontaneous tumours. BMC Cancer. 2008;8:2.
29. Ferretti S, Allegrini PR, O'Reilly T, Schnell C, Stumm M, Wartmann M, et al. Patupilone induced vascular disruption in orthotopic rodent tumor models detected by magnetic resonance imaging and interstitial fluid pressure. Clin Cancer Res. 2005;11(21):7773–84.
30. Ferretti S, Allegrini PR, Becquet MM, McSheehy PM. Tumor interstitial fluid pressure as an early-response marker for anticancer therapeutics. Neoplasia. 2009;11(9):874–81.

31. Boucher Y, Kirkwood JM, Opacic D, Desantis M, Jain RK. Interstitial hypertension in superficial metastatic melanomas in humans. Cancer Res. 1991;51(24):6691–4.

32. Shieh AC. Biomechanical forces shape the tumor microenvironment. Ann Biomed Eng. 2011;39(5):1379–89.

33. Venkatesh S, Wang G, Madhavan A, Wee A, editors. Comparison of MR elastography and diffusion weighted MR imaging for differentiating benign and malignant focal liver lesions. In: Proceedings of ECR 2012 (abstract B-0827). European Congress of Radiology; 2012; Vienna, Austria: ESR.

34. Li J, Jamin Y, Boult JK, Cummings C, Waterton JC, Ulloa J, et al. Tumour biomechanical response to the vascular disrupting agent ZD6126 in vivo assessed by magnetic resonance elastography. Br J Cancer. 2014;110(7):1727–32.

35. Motosugi U, Ichikawa T, Koshiishi T, Sano K, Morisaka H, Ichikawa S, et al. Liver stiffness measured by magnetic resonance elastography as a risk factor for hepatocellular carcinoma: a preliminary case-control study. Eur Radiol. 2013;23(1):156–62.

36. Nahon P, Kettaneh A, Lemoine M, Seror O, Barget N, Trinchet JC, et al. Liver stiffness measurement in patients with cirrhosis and hepatocellular carcinoma: a case-control study. Eur J Gastroenterol Hepatol. 2009;21(2):214–9.

37. Anaparthy R, Talwalkar JA, Yin M, Roberts LR, Fidler JL, Ehman RL. Liver stiffness measurement by magnetic resonance elastography is not associated with developing hepatocellular carcinoma in subjects with compensated cirrhosis. Aliment Pharmacol Ther. 2011;34:83–91.

38. Mariappan YK, Glaser KJ, Manduca A, Romano AJ, Venkatesh SK, Yin M, et al. High-frequency mode conversion technique for stiff lesion detection with magnetic resonance elastography (MRE). Magn Reson Med. 2009;62(6):1457–65.

Impact of Magnetic Resonance Elastography on Liver Diseases: Clinical Perspectives

Sumeet K. Asrani and Jayant A. Talwalkar

Introduction

Burden of Chronic Liver Disease

Chronic liver disease and cirrhosis remains a major public health problem worldwide. In the United States alone, an estimated 150,000 persons annually are diagnosed with chronic liver disease with nearly 30,000 (20 %) individuals having cirrhosis at initial presentation. Furthermore, over 75 % of individuals are likely to be asymptomatic from their liver disease at diagnosis [1]. These trends are expected to increase based on an aging population and the growing epidemic of obesity [2, 3].

Current diagnostic tests used within clinical practice are not sensitive or specific enough to function as screening tests for more invasive, definitive procedures such as liver biopsy. Although liver biopsy has been considered the gold standard for detecting hepatic fibrosis, its use in clinical practice appears to be declining over time. The availability of laboratory assays to identify liver diseases such as chronic hepatitis C is a major reason for this phenomenon. Additionally, the ability to perform liver biopsy in large populations are limited by the risk for procedure-related complications, patient acceptance, cost, and inaccuracies associated with sampling error [4, 5]. In turn, there remains a need to introduce novel, effective methods to diagnose and risk stratify individuals with chronic liver disease.

Contemporary Issues with Liver Biopsy

Liver histology classification systems recognize five stages of fibrosis, defined as F0 (no fibrosis), F1 (portal fibrosis), F2 (periportal fibrosis), F3 (bridging fibrosis), and F4 (cirrhosis). Clinically significant hepatic fibrosis is defined by the presence of periportal fibrosis (stage F2) or greater on liver histology. While measurement error with liver biopsy has been an acknowledged limitation for decades, recent studies have systematically demonstrated that a minimum liver tissue size (≥ 20 mm) and portal tract number (≥ 10) are needed to maximize the accuracy of biopsy [4, 5]. In clinical practice, however, the average liver tissue size and portal tract number are often less than this threshold [6]. Hence, the ability to make reliable and valid fibrosis staging assessments remains inexact for a significant number of patients.

S.K. Asrani, M.D., M.Sc.
Baylor University Medical Center, 3410 Worth Street, Suite 860, Dallas, TX 75246, USA
e-mail: Sumeet.asrani@baylorhealth.edu

J.A. Talwalkar, M.D., M.P.H. (✉)
Division of Gastroenterology and Hepatology, Mayo Clinic, 200 First Street SW, Rochester, MN 55905, USA
e-mail: talwalkar.jayant@mayo.edu

S.K. Venkatesh and R.L. Ehman (eds.), *Magnetic Resonance Elastography*,
DOI 10.1007/978-1-4939-1575-0_6, © Springer Science+Business Media New York 2014

Elastography Imaging for Detecting Hepatic Fibrosis

Given the practical limitations of using liver biopsy among the growing number of individuals with chronic liver disease, a number of noninvasive methods to detect liver fibrosis have been developed over the past decade. Furthermore, the ability to characterize fibrosis extent by categories as opposed to discrete stages has gained importance as more effective therapies for chronic liver disease become available. In general, these methods have been characterized as (1) serum tests containing panels of markers reflecting the biology of hepatic fibrogenesis and (2) functional imaging tests which measure unique physical characteristics associated with the development of fibrosis. Among the imaging studies to date, the most widely studied and used are elastography imaging techniques with ultrasound or magnetic resonance (MR) platforms [7].

Ultrasound-based transient elastography (TE) utilizes a transducer probe which emits low-frequency (50 Hz) vibrations into the liver when applied between the rib interspaces where biopsy would be performed. The propagating shear wave induced by these vibrations is then detected by a pulse-echo acquisition, and the velocity of the wave is then calculated. Shear wave velocity is measured in kiloPascals (kPa), and is proportional to liver stiffness represented by the equation for Young's elastic modulus E (expressed as $E = 3\rho v^2$, where v is the shear velocity and ρ is the density of tissue, assumed to be constant). The accurate measurement of liver stiffness by TE is represented by (1) the ratio of successful measurements to the total number of acquisitions being ≥ 60 % and (2) the interquartile range for measurements being within 30 % of the median value [8].

Magnetic resonance elastography (MRE) uses a modified phase-contrast imaging sequence to detect propagating shear waves within the tissue of interest. Acoustic shear waves are generated by a pneumatic driver placed directly over the upper abdomen for propagation into liver tissue. Subsequently, the liver stiffness values are calculated from wave displacement patterns and displayed as color-encoded images (elastograms). Region of interest analysis throughout four cross-sectional slices of liver (avoiding vascular structures) is then performed to calculate mean liver stiffness [9, 10]. Elasticity quantification by MRE is based on the formula representing shear modulus, which is equivalent to one third of the Young's modulus which is used with TE [11].

Clinical Impact of Elastography Imaging on Liver Disease

A major goal within clinical practice is to identify patients who have developed clinically significant hepatic fibrosis or cirrhosis. Decisions regarding monitoring versus institution of disease-specific therapy are now dictated by the category of fibrosis involvement rather than its specific individual stage. For most noninvasive tests, the ability to categorize patients as having no to minimal fibrosis (F0–F1) or moderate to severe fibrosis (F2–F4) is now possible [7, 8]. Studies to date have shown that liver stiffness assessment by MRE and TE is highly effective at separating individual patients into these diagnostic categories. Liver stiffness as measured by MRE is highly reproducible with high interobserver agreement [12, 13]. Further it may also be feasible and highly accurate among the pediatric population, including those with obesity [14].

Detection of Clinically Significant Hepatic Fibrosis and Cirrhosis

The first clinical application of MRE has been the detection and characterization of hepatic fibrosis. In this diagnostic role, MRE offers a safer and potentially more accurate alternative to invasive liver biopsy. Pilot and initial prospective studies have demonstrated the feasibility and diagnostic accuracy in detecting hepatic fibrosis among patients with known chronic liver disease [15–19]. MRE has been shown to be highly effective for distinguishing normal from fibrotic livers with a very high negative predictive value exceeding 97 %. Receiver operating characteristic (ROC) analysis has demonstrated that a liver stiffness value ≥ 2.9 kPa separates any degree of liver fibrosis from normal liver with a sensitivity and specificity of 98 and 99 %, respectively.

Conversely, MRE is highly accurate for the detection of cirrhosis with sensitivity and specificity values exceeding 90 %, respectively [16, 17]. For example, among persons with chronic hepatitis B, MRE was significantly more accurate than serum fibrosis markers for the detection of significant fibrosis (0.99 vs. 0.55–0.73) and cirrhosis (0.98 vs. 0.53–0.77). Sensitivity, specificity, positive predictive, and negative predictive values for MRE for cirrhosis were 97.4, 100, 100, and 96, and 100, 95.2, 91.3, and 100 %, respectively [20].

This compares favorably with TE where average sensitivity and specificity rates nearing 90 %, respectively, are also observed [21, 22]. The importance of identifying cirrhosis at its earliest stages is for detecting complications such as esophageal varices and hepatocellular carcinoma, both of which can benefit from early interventions.

Furthermore, MRE can discriminate between patients with clinically significant hepatic fibrosis (F2–F4) as compared to individuals with no to mild fibrosis (F0–F1) with sensitivity and specificity values in the 80–85 % range, respectively [10, 17]. Notably, these results are higher when compared to TE where sensitivity and specificity values between 70 and 80 %, respectively, have been reported in single and combined analyses [21, 22]

Recently, a large prospective study comparing MRE, TE, and the serum AST-to-platelet ratio index (APRI) in a series of 141 patients with chronic liver disease was performed [17]. In addition to a higher technical success rate for MRE vs. TE (94 vs. 84 %), the diagnostic accuracy of MRE was also superior to both TE and APRI. While comparable accuracy rates for detecting cirrhosis were observed for MRE and TE, the accuracy of MRE was significantly higher than TE for detecting clinically significant hepatic fibrosis (F2–F4).

The comparative study between MRE and TE highlights known inherent differences between these techniques in assessing hepatic fibrosis. While the reproducibility of TE is excellent in experienced centers, its accuracy is diminished by the presence of inexperienced operators and patient factors including obesity and narrow rib interspaces [23, 24]. This is based, in part, on the inability of TE to provide wave penetration beyond a distance of 6–7 cm which can represent the width of subcutaneous tissues in some obese patients [23, 24]. Thus, a major issue for TE is the prevalence and severity of obesity in developed countries, where the frequency of incomplete exam and complete technical failure rates may exceed 30 % [23, 24]. The reproducibility of MRE is also excellent [25]. In a recent study comparing MRE and shear wave elastography in the same individual, the rates of unreliable exam were higher with shear wave as compared to MRE (19 vs. 0 %) [26]. Furthermore, MRE is not significantly affected by obesity, rib interspace width, or ascites. While TE is designed to measure liver stiffness in a cylindrical-shaped area 1 cm wide and 4 cm long confined exclusively to the peripheral right liver, MRE calculates liver stiffness over the entire cross-sectional areas of hepatic parenchyma from multiple slices. Individuals with typical contraindications to MRI, however, are unable to undergo MRE [11, 24].

For both MRE and TE, it should also be noted that other pathophysiological processes including acute inflammation, cholestasis, portal pressure, and hepatic congestion may independently contribute to liver stiffness [27–31]. A recent study evaluated the role of MRE in assessment of liver fibrosis among patients with congenital heart disease after the Fontan procedure. In this preliminary study, the mean liver stiffness was elevated at levels associated with advanced fibrosis. However, it is unclear whether in this population the elevated stiffness is marker of increased passive congestion or a reflection of true fibrosis due to longstanding cardiac dysfunction [32]. This emphasizes the point that elevated liver stiffness may have different implications depending upon the population.

On the other had it is noted that mean liver stiffness among healthy controls is not influenced by age, gender, liver fat content, or body mass index [33, 34].

Detection of Clinically Significant Portal Hypertension

Complications associated with portal hypertension represent a significant proportion of morbidity and mortality among individuals with cirrhosis. The gold standard test for detecting portal hypertension involves measurement of the hepatic venous pressure gradient (HVPG) by transjugular hepatic venography. Numerous studies to date have identified an HVPG $\geq$ 10 mmHg as an independent predictor of developing complications of liver disease. Furthermore, a reduction in HVPG below 12 mmHg (or HVPG reduction by $\geq$20 %) with treatment after variceal bleeding is associated with improved survival [35]. However, the measurement of HVPG in clinical practice has not become widespread based on the technical demands, invasiveness, and risk for complications associated with this procedure [35, 36].

Given the ability of elastography imaging to detect cirrhosis, a number of subsequent investigations have demonstrated that liver stiffness is strongly correlated with HVPG values below 10 mmHg among individuals with chronic hepatitis C [37–39]. The correlation is less precise, however, beyond an HVPG value of 12 mmHg based on complex pathophysiologic changes associated with advanced portal hypertension such as increasing resistance to portal blood flow, porto-systemic collaterals, and splanchnic vasodilatation [40, 41]. The potential use of liver stiffness for detecting esophageal varices in patients has also been studied. However, these have been studies using TE which show that liver stiffness is an insensitive predictor for esophageal varices among individuals with cirrhosis [40, 42]. Similar results have also been observed with MRE [43]. At this point, it cannot be recommended that elastography imaging be used as a screening method to determine which patients with compensated cirrhosis should undergo endoscopic screening for esophageal varices. On the other hand, a recent study did show promise in the prediction of presence of large varices (as compared to small varices). In this single center study, a cutoff of 5.8 kPA had good sensitivity (96 %) and negative predictive value (98 %) for prediction of large varices though the specificity (60 %) and positive predictive value (36 %) was poor [44].

Given the relationship between splenomegaly and esophageal varices with cirrhosis, spleen stiffness as a noninvasive predictor of portal hypertension was recently measured using MRE [43]. In addition to demonstrating that spleen stiffness was significantly higher among patients with liver fibrosis compared to healthy controls, the spleen stiffness was increased further among patients with cirrhosis and porto-systemic collaterals versus patients with cirrhosis alone. Notably, a mean spleen stiffness $\geq$10.5 kPa was 100 % sensitive in identifying all 17 patients with esophageal varices and cirrhosis in the study cohort. Therefore, it is possible that spleen stiffness provides additional information about the hemodynamic alterations within splenic and splanchnic circulations not captured by liver stiffness. Based on these preliminary results, future studies including the measurement of HVPG as well as spleen and liver stiffness are being pursued.

Compensated and Decompensated Cirrhosis

Patients affected by compensated and decompensated (presence of variceal bleeding, ascites or hepatic encephalopathy) cirrhosis are known to have disparate clinical outcomes. As compared to the general population, individuals with compensated cirrhosis have a fivefold increase, whereas patients with decompensated disease have a tenfold increase in mortality [45]. Given that a majority of deaths in patients with compensated cirrhosis are due to progression to a decompensated state and the development of its ensuing complications, the ability to predict decompensation is important. If patients with compensated liver disease at the highest risk of decompensation can be identified, it may be possible to institute enhanced surveillance and prophylactic measures for this patient subset. The role of elastography in identifying this risk of transition has been reported. In a recent meta-analysis, the degree of liver stiffness as quantified by either TE or MRE was associated with increased risk of decompensated cirrhosis, HCC, and death [46]. Specifically, in a single center study, subjects with decompensated cirrhosis had a higher mean liver stiffness (6.8 vs. 5.2 kPa). MRE was independently associated with decompensation after adjusting for relevant patient characteristics. Further among persons with compensated cirrhosis, the hazard for hepatic decompensation was 1.42 (95 % CI 1.16–1.75) per unit increase in liver shear stiffness over time. The hazard of hepatic decompensation was 4.96 (95 % CI 1.4–17.0, $p=0.019$) for a subject with compensated disease and mean liver stiffness value ≥ 5.8 kPa as compared to an individual with compensated disease and lower mean liver stiffness values [47].

Other exciting yet novel applications of MRE are also currently being evaluated. This includes assessment of the role of elastography in prediction of decompensated disease after liver transplantation especially among persons with hepatitis C [48–50].

Detection of Clinically Significant Focal Hepatic Lesions

With the growth in medical imaging, an increasing number of individuals are being identified with incidental focal hepatic lesions. For most lesions, conventional imaging can identify radiographic features of several benign lesions as well as those concerning for malignancy. Some lesions, however, remain indeterminate in nature despite multiple imaging studies. Recently, the technique of liver MRE was applied to characterize liver stiffness among a variety of benign and malignant lesions [51]. Among 64 patients with 44 hepatic mass lesions, the mean stiffness of benign masses (9 hemangiomas, 3 focal nodular hyperplasias, and 1 hepatic adenoma) was 2.7 kPa, which is consistent with the liver stiffness of normal liver parenchyma. The mean stiffness of malignant tumors was 10.1 kPa. In this series, a cutoff value of 5 kPa completely separated all benign focal liver masses from malignant lesions. In a case control study, liver stiffness as quantified by MRE was an independent risk factor for HCC (odds ratio 1.38 (1.05–1.84)) [52]. However, other studies have failed to show this association [53].

In current practice, an immediate potential utility for MRE is to improve the characterization of benign lesions which are atypical in presentation, especially when the distinction between atypical cavernous hemangioma, focal nodular hyperplasia, or hepatic adenoma from malignant disease is not clear from conventional imaging techniques [54]. However, the use of contrast-enhanced techniques has been highly accurate for the diagnosis of hepatocellular carcinoma in patients with cirrhosis when lesions are at least 2 cm in size, contain arterial hypervascularity, and have a delayed venous washout pattern [55]. While the application of MRE for hepatocellular carcinoma does not seem to provide obvious incremental diagnostic value, it remains to be seen whether tumor stiffness can predict responses to non-surgical therapies including ablative procedures and pharmacological therapies. Furthermore, the measurement of liver stiffness after chemotherapy for colorectal hepatic metastases may help to identify individuals who may be at risk for hepatic sinusoidal injury.

References

1. Bell BP, Manos MM, Zaman A, Terrault N, Thomas A, Navarro VJ, et al. The epidemiology of newly diagnosed chronic liver disease in gastroenterology practices in the United States: results from population-based surveillance. Am J Gastroenterol. 2008; 103(11):2727–36. quiz 37.
2. Liu B, Balkwill A, Reeves G, Beral V. Body mass index and risk of liver cirrhosis in middle aged UK women: prospective study. BMJ. 2010;340:c912.
3. Kim WR, Brown Jr RS, Terrault NA, El-Serag H. Burden of liver disease in the United States: summary of a workshop. Hepatology. 2002;36(1):227–42.
4. Muir AJ, Trotter JF. A survey of current liver biopsy practice patterns. J Clin Gastroenterol. 2002;35(1): 86–8.
5. Standish RA, Cholongitas E, Dhillon A, Burroughs AK, Dhillon AP. An appraisal of the histopathological assessment of liver fibrosis. Gut. 2006;55(4):569–78.
6. Everhart JE, Wright EC, Goodman ZD, Dienstag JL, Hoefs JC, Kleiner DE, et al. Prognostic value of Ishak fibrosis stage: findings from the hepatitis C antiviral long-term treatment against cirrhosis trial. Hepatology. 2010;51(2):585–94.
7. Castera L, Pinzani M. Non-invasive assessment of liver fibrosis: are we ready? Lancet. 2010; 375(9724):1419–20.
8. Castera L, Forns X, Alberti A. Non-invasive evaluation of liver fibrosis using transient elastography. J Hepatol. 2008;48(5):835–47.

9. Muthupillai R, Lomas DJ, Rossman PJ, Greenleaf JF, Manduca A, Ehman RL. Magnetic resonance elastography by direct visualization of propagating acoustic strain waves. Science. 1995;269(5232):1854–7.

10. Yin M, Chen J, Glaser KJ, Talwalkar JA, Ehman RL. Abdominal magnetic resonance elastography. Top Magn Reson Imaging. 2009;20(2):79–87.

11. Talwalkar JA, Yin M, Fidler JL, Sanderson SO, Kamath PS, Ehman RL. Magnetic resonance imaging of hepatic fibrosis: emerging clinical applications. Hepatology. 2008;47(1):332–42.

12. Lee YJ, Lee JM, Lee JE, Lee KB, Lee ES, Yoon JH, et al. MR elastography for noninvasive assessment of hepatic fibrosis: reproducibility of the examination and reproducibility and repeatability of the liver stiffness value measurement. J Magn Reson Imaging. 2014;39(2):326–31.

13. Runge JH, Bohte AE, Verheij J, Terpstra V, Nederveen AJ, van Nieuwkerk KM, et al. Comparison of interobserver agreement of magnetic resonance elastography with histopathological staging of liver fibrosis. Abdom Imaging. 2013;39(2):283–90.

14. Xanthakos SA, Podberesky DJ, Serai SD, Miles L, King EC, Balistreri WF, et al. Use of magnetic resonance elastography to assess hepatic fibrosis in children with chronic liver disease. J Pediatr. 2014;164(1):186–8.

15. Rouviere O, Yin M, Dresner MA, Rossman PJ, Burgart LJ, Fidler JL, et al. MR elastography of the liver: preliminary results. Radiology. 2006;240(2):440–8.

16. Yin M, Talwalkar JA, Glaser KJ, Manduca A, Grimm RC, Rossman PJ, et al. Assessment of hepatic fibrosis with magnetic resonance elastography. Clin Gastroenterol Hepatol. 2007;5(10):1207–13 e2.

17. Huwart L, Sempoux C, Salameh N, Jamart J, Annet L, Sinkus R, et al. Liver fibrosis: noninvasive assessment with MR elastography versus aspartate aminotransferase-to-platelet ratio index. Radiology. 2007;245(2):458–66.

18. Asbach P, Klatt D, Hamhaber U, Braun J, Somasundaram R, Hamm B, et al. Assessment of liver viscoelasticity using multifrequency MR elastography. Magn Reson Med. 2008;60(2):373–9.

19. Crespo S, Bridges M, Nakhleh R, McPhail A, Pungpapong S, Keaveny AP. Non-invasive assessment of liver fibrosis using magnetic resonance elastography in liver transplant recipients with hepatitis C. Clin Transplant. 2013;27(5):652–8.

20. Venkatesh SK, Wang G, Lim SG, Wee A. Magnetic resonance elastography for the detection and staging of liver fibrosis in chronic hepatitis B. Eur Radiol. 2014;24(1):70–8.

21. Talwalkar JA, Kurtz DM, Schoenleber SJ, West CP, Montori VM. Ultrasound-based transient elastography for the detection of hepatic fibrosis: systematic review and meta-analysis. Clin Gastroenterol Hepatol. 2007;5(10):1214–20.

22. Friedrich-Rust M, Ong MF, Martens S, Sarrazin C, Bojunga J, Zeuzem S, et al. Performance of transient elastography for the staging of liver fibrosis: a meta-analysis. Gastroenterology. 2008;134(4):960–74.

23. Castera L, Foucher J, Bernard PH, Carvalho F, Allaix D, Merrouche W, et al. Pitfalls of liver stiffness measurement: a 5-year prospective study of 13,369 examinations. Hepatology. 2010;51(3):828–35.

24. Talwalkar JA. Elastography for detecting hepatic fibrosis: options and considerations. Gastroenterology. 2008;135(1):299–302.

25. Hines CD, Bley TA, Lindstrom MJ, Reeder SB. Repeatability of magnetic resonance elastography for quantification of hepatic stiffness. J Magn Reson Imaging. 2010;31(3):725–31.

26. Yoon JH, Lee JM, Woo HS, Yu MH, Joo I, Lee ES, et al. Staging of hepatic fibrosis: comparison of magnetic resonance elastography and shear wave elastography in the same individuals. Korean J Radiol. 2013;14(2):202–12.

27. Arena U, Vizzutti F, Corti G, Ambu S, Stasi C, Bresci S, et al. Acute viral hepatitis increases liver stiffness values measured by transient elastography. Hepatology. 2008;47(2):380–4.

28. Sagir A, Erhardt A, Schmitt M, Haussinger D. Transient elastography is unreliable for detection of cirrhosis in patients with acute liver damage. Hepatology. 2008;47(2):592–5.

29. Millonig G, Reimann FM, Friedrich S, Fonouni H, Mehrabi A, Buchler MW, et al. Extrahepatic cholestasis increases liver stiffness (FibroScan) irrespective of fibrosis. Hepatology. 2008;48(5):1718–23.

30. Millonig G, Friedrich S, Adolf S, Fonouni H, Golriz M, Mehrabi A, et al. Liver stiffness is directly influenced by central venous pressure. J Hepatol. 2010;52(2):206–10.

31. Chang PE, Miquel R, Blanco JL, Laguno M, Bruguera M, Abraldes JG, et al. Idiopathic portal hypertension in patients with HIV infection treated with highly active antiretroviral therapy. Am J Gastroenterol. 2009;104(7):1707–14.

32. Serai SD, Wallihan DB, Venkatesh SK, Ehman RL, Campbell KM, Sticka J, et al. Magnetic resonance elastography of the liver in patients status-post fontan procedure: feasibility and preliminary results. Congenit Heart Dis. 2014;9(1):7–14.

33. Venkatesh SK, Wang G, Teo LL, Ang BW. Magnetic resonance elastography of liver in healthy Asians: normal liver stiffness quantification and reproducibility assessment. J Magn Reson Imaging. 2014;39(1):1–8.

34. Lee DH, Lee JM, Han JK, Choi BI. MR elastography of healthy liver parenchyma: normal value and reliability of the liver stiffness value measurement. J Magn Reson Imaging. 2013;38(5):1215–23.

35. Burroughs AK, Thalheimer U. Hepatic venous pressure gradient in 2010: optimal measurement is key. Hepatology. 2010;51(6):1894–6.

36. Bureau C, Metivier S, Peron JM, Selves J, Robic MA, Gourraud PA, et al. Transient elastography accurately predicts presence of significant portal hypertension in patients with chronic liver disease. Aliment Pharmacol Ther. 2008;27(12):1261–8.

37. Groszmann RJ, Wongcharatrawee S. The hepatic venous pressure gradient: anything worth doing should be done right. Hepatology. 2004;39(2):280–2.

38. Lemoine M, Katsahian S, Ziol M, Nahon P, Ganne-Carrie N, Kazemi F, et al. Liver stiffness measurement as a predictive tool of clinically significant portal hypertension in patients with compensated hepatitis C virus or alcohol-related cirrhosis. Aliment Pharmacol Ther. 2008;28(9):1102–10.

39. Vizzutti F, Arena U, Romanelli RG, Rega L, Foschi M, Colagrande S, et al. Liver stiffness measurement predicts severe portal hypertension in patients with HCV-related cirrhosis. Hepatology. 2007;45(5):1290–7.

40. Lim JK, Groszmann RJ. Transient elastography for diagnosis of portal hypertension in liver cirrhosis: is there still a role for hepatic venous pressure gradient measurement? Hepatology. 2007;45(5):1087–90.

41. Castera L, Le Bail B, Roudot-Thoraval F, Bernard PH, Foucher J, Merrouche W, et al. Early detection in routine clinical practice of cirrhosis and oesophageal varices in chronic hepatitis C: comparison of transient elastography (FibroScan) with standard laboratory tests and non-invasive scores. J Hepatol. 2009;50(1):59–68.

42. Kazemi F, Kettaneh A, N'Kontchou G, Pinto E, Ganne-Carrie N, Trinchet JC, et al. Liver stiffness measurement selects patients with cirrhosis at risk of bearing large oesophageal varices. J Hepatol. 2006;45(2):230–5.

43. Talwalkar JA, Yin M, Venkatesh S, Rossman PJ, Grimm RC, Manduca A, et al. Feasibility of in vivo MR elastographic splenic stiffness measurements in the assessment of portal hypertension. AJR Am J Roentgenol. 2009;193(1):122–7.

44. Sun HY, Lee JM, Han JK, Choi BI. Usefulness of MR elastography for predicting esophageal varices in cirrhotic patients. J Magn Reson Imaging. 2013;39(3):559–66.

45. Fleming KM, Aithal GP, Card TR, West J. All-cause mortality in people with cirrhosis compared with the general population: a population-based cohort study. Liver Int. 2012;32(1):79–84.

46. Singh S, Fujii LL, Murad MH, Wang Z, Asrani SK, Ehman RL, et al. Liver stiffness is associated with risk of decompensation, liver cancer, and death in patients with chronic liver diseases: a systematic review and meta-analysis. Clin Gastroenterol Hepatol. 2013;11(12):1573–84 e1-2. quiz e88-9.

47. Asrani SK, Talwalkar JA, Kamath PS, Shah VH, Saracino G, Jennings L, et al. Role of magnetic resonance elastography in compensated and decompensated liver disease. J Hepatol. 2013;60(5):934–9.

48. Crespo G, Lens S, Gambato M, Carrion JA, Marino Z, Londono MC, et al. Liver stiffness 1 year after transplantation predicts clinical outcomes in patients with recurrent hepatitis C. Am J Transplant. 2014;14(2):375–83.

49. Kamphues C, Klatt D, Bova R, Yahyazadeh A, Bahra M, Braun J, et al. Viscoelasticity-based magnetic resonance elastography for the assessment of liver fibrosis in hepatitis C patients after liver transplantation. RoFo. 2012;184(11):1013–9.

50. Lee VS, Miller FH, Omary RA, Wang Y, Ganger DR, Wang E, et al. Magnetic resonance elastography and biomarkers to assess fibrosis from recurrent hepatitis C in liver transplant recipients. Transplantation. 2011;92(5):581–6.

51. Venkatesh SK, Yin M, Glockner JF, Takahashi N, Araoz PA, Talwalkar JA, et al. MR elastography of liver tumors: preliminary results. AJR Am J Roentgenol. 2008;190(6):1534–40.

52. Motosugi U, Ichikawa T, Koshiishi T, Sano K, Morisaka H, Ichikawa S, et al. Liver stiffness measured by magnetic resonance elastography as a risk factor for hepatocellular carcinoma: a preliminary case-control study. Eur Radiol. 2013;23(1):156–62.

53. Anaparthy R, Talwalkar JA, Yin M, Roberts LR, Fidler JL, Ehman RL. Liver stiffness measurement by magnetic resonance elastography is not associated with developing hepatocellular carcinoma in subjects with compensated cirrhosis. Aliment Pharmacol Ther. 2011;34(1):83–91.

54. Garteiser P, Doblas S, Daire JL, Wagner M, Leitao H, Vilgrain V, et al. MR elastography of liver tumours: value of viscoelastic properties for tumour characterisation. Eur Radiol. 2012;22(10):2169–77.

55. Oki E, Kakeji Y, Taketomi A, Yamashita Y, Ohgaki K, Harada N, et al. Transient elastography for the prediction of oxaliplatin-associated liver injury in colon cancer patients: a preliminary analysis. J Gastrointest Cancer. 2008;39(1–4):82–5.

Magnetic Resonance Elastography of the Skeletal Muscle

Sabine F. Bensamoun

Introduction

Muscle tissue can be evaluated with both invasive and non-invasive techniques. Invasive methods include muscle biopsy and electromyography. Muscle biopsy determines the muscle composition while electromyography using needles can characterize the muscle contractile efficiency. Non-invasive methods include palpation and imaging techniques such as ultrasound (US) and magnetic resonance imaging (MRI) which are useful for the quantification of the morphological properties (muscle volume, cross sectional area, pennation angle). In addition to the structural parameters, it is also important to characterize the mechanical properties of the muscle, which acts as an active tissue composed of contractile (myosin) and elastic passive (actin, connective tissue) components. Magnetic Resonance Elastography (MRE) is capable of quantifying the muscle stiffness [1–3] from the displacement of shear waves inside the muscle and imaging its entire spatial distribution. Several studies [4–6] have developed and applied MRE technique to the skeletal muscle and reference values of the muscle stiffness have

been established in order to provide clinicians quantitative stiffness measurement, allowing the differentiation of the healthy and pathological muscles based on stiffness values [7–9]. MRE may provide a better understanding of the structural and functional properties of muscle following treatment and therapy, and thus help prevent muscle disorders.

MRE of Muscle Technique

As the muscle is an active tissue, mechanical properties would be best assessed in both resting and active states. The MRE technique therefore needs additional experimental set up to assess mechanical properties under load conditions. The MRE technique described here is with thigh musculature as an example. The technique remains same for other skeletal muscles and body regions with suitable modifications for loaded conditions.

To investigate the mechanical properties of the thigh muscle, a MR compatible leg press was developed to measure the muscle stiffness in a relaxed and contracted condition [10 and 20 % of the maximum voluntary contraction (MVC)]. The volunteers lay down on the leg press with the shoulders maintained in a fixed position inside a 1.5 T MRI scanner (General Electric HDxt). The knee is flexed at 30° and the right foot is placed on a footplate and secured with Velcro straps. The footplate is composed of a MR compatible

S.F. Bensamoun, Ph.D. (✉)
Laboratoire de Biomécanique et Bioingénierie,
UMR CNRS 7338, Université de Technologie de
Compiègne (UTC), Compiègne, France
e-mail: sabine.bensamoun@utc.fr;
sabine.bensamoun@gmail.com

S.K. Venkatesh and R.L. Ehman (eds.), *Magnetic Resonance Elastography*,
DOI 10.1007/978-1-4939-1575-0_7, © Springer Science+Business Media New York 2014

load cell to measure the developed force during the isometric contractions. A custom-made LABVIEW program records the developed forces and gives a visual feedback to the volunteer to ensure the desired force was maintained and reached for the time of the scan.

Different types of driver (mechanical or pneumatic) can be used to generate shear waves inside the thigh muscles [1]. The mechanical driver used to investigate muscles located near skin surface and without thick adipose tissue above them. The pneumatic driver is useful to generate shear waves in deeply seated muscles. Thigh muscles are usually surrounded by subcutaneous tissue requiring the use of a pneumatic driver. This driver consists of a remote pressure source (i.e., a large active loudspeaker) connected to a long plastic tube similar to the MRE of liver technique described in Chap. 3. A smaller silicone tube is wrapped and clamped on the thigh, one-third of the distance from the patellar tendon to the greater trochanter, and is connected to the remote pressure driver. This system will create a time varying pressure wave, which will cause the tube around the thigh to expand or contract with the remote driver at 90 Hz (f). A custom-made Helmholtz surface receive coil is also placed around the thigh for data acquisition.

MRE Sequence

A gradient echo (GRE) MRI sequence is used to obtain axial images of the thigh muscles where oblique scan planes for MRE acquisitions are prescribed though the investigated muscle. Then, MRE images reveal the target muscles where the propagation of the shear waves will be analyzed. MRE is performed with a 256×64 acquisition matrix (interpolated to 256×256), a flip angle of 45°, and a 24 cm field of view. Four offsets were recorded for each muscle in relaxed and contracted (10, 20 % MVC) states. The scan time is 52 s using a TR/TE of 100 ms/minimum full.

MRE technique provides anatomical images of the vastus medialis (VM) muscle (Fig. 7.1a) as well as phase images (Fig. 7.1b) showing the shear wave displacement within the muscle. A black profile is manually placed in the direction of the wave propagation allowing the quantification of the wavelength (λ) that leads to the measurement of the shear modulus ($\mu = \rho\lambda^2 f^2$, with $\rho = 1,000$ kg/m³ for the muscle density) assuming that the muscle is linearly elastic, locally homogeneous, isotropic, and incompressible.

To further characterize the micro structural changes occurring in muscle tissue, the viscoelastic (μ, η) properties of the muscles as well as the subcutaneous adipose tissue can be determined with multifrequency MRE and different rheological models (Voigt, Zener, and springpot) [10].

In addition, the attenuation coefficient (α) is calculated from the shear wave displacement amplitude in function of the distance. An exponential fitting curve.

$Ae^{-(\alpha d)}$ is used to determine the attenuation coefficient, with A and d representing the displacement amplitude value and the distance along the profile, respectively.

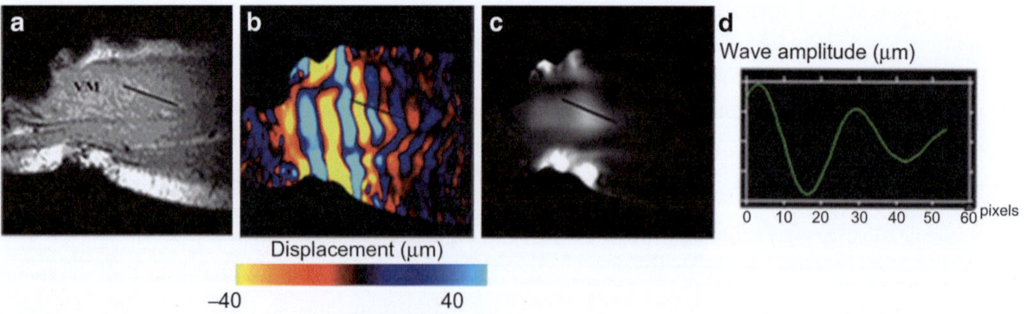

Fig. 7.1 **a**: Anatomical image of the vastus medialis (VM) muscle. **b**: Phase image with a black profile drawn along the propagation of the shear waves. **c**: Amplitude image resulting from the amplitude value of the first temporal harmonic extracted at the driven frequency 90 Hz along each pixel. **d**: Shear wave along the profile

Fig. 7.2 Cartography of the muscle stiffness

A map of the shear modulus (Fig. 7.2) is generated from the wave displacement image using the Linear Frequency Estimate (LFE) algorithm [11] providing a spatial distribution of the muscle tissue stiffness. Assuming that muscle tissue is locally homogeneous, only the region of interest (ROI) around the prescribed profile may be analyzed.

Experimental Setup for Ultrasound Imaging

For comparative evaluation ultrasound can also be performed. Ultrasonic images are performed with B-mode using a standard ultrasound machine with suitable probe on the subject's thigh who lays supine on the same previous leg press and with the same position as for the MRE test [12]. To visualize the anatomical structure located in the same area as the shear wave's propagation, the tube wrapped around the thigh is used as a reference in order to localize its indentation observed on the MRI anatomical image (Fig. 7.3a). Then, the ultrasound transducer is placed perpendicularly and moved around the tube until the visualization of the vastus medialis (VM) and sartorius (Sr) muscles (Fig. 7.3b). Marks are done with a pen on the skin, to identify the placement of the probe which is the same for

all experiments. Longitudinal ultrasonic images (resolution 960×720 mm) showing the parallel-oriented pattern of fascicles are recorded in both relaxed and contracted (10 and 20 % MVC) conditions for each subject. Wave angle (α_MRE, Fig. 7.3a) and ultrasound fascicle angles (α_US, Fig. 7.3b) are calculated between the subcutaneous adipose tissue-muscle interface and the shear wave propagation, and the path of the fascicle, respectively. The accuracy of the measurement is $0.5°$ and the reproducibility is about 6 %.

Database of Thigh Muscle Stiffness Using MRE

A stiffness database of the vastus lateralis (VL), vastus medialis (VM), and sartorius (Sr) muscles has been established for adults volunteers (mean age$=25.2 \pm 1.78$) and the shear modulus measured at rest for VL, VM, and Sr are 3.73 ± 0.85, 3.91 ± 1.15, and 7.53 ± 1.63 kPa, respectively [1]. The higher stiffness found for the sartorius indicates that the propagation of the waves is related to the muscle architecture or orientation of the muscle fibers. Indeed, vasti's muscles are composed of oblique muscle fibers while the sartorius has longitudinally oriented muscle fibers. In contraction, the stiffness of both vasti increased with the level of contraction, compared to the sartorius stiffness which remained the same, due to the applied isometric contraction which solicits only the knee extensors while the sartorius flexes the knee. Bensamoun's study (2006) has demonstrated that MRE technique is able to determine the stiffness of different thigh muscles reflecting the muscle architecture (unipennate, longitudinal) and the muscle fiber composition (type I, II).

Effects of Muscle Conditions (Relaxed and Contracted) and Age on the Shear Modulus and Attenuation Coefficient

Analysis of the shear modulus (Fig. 7.4a) showed progressive muscle fibers recruitment with the degree of contraction for young adults (24-29 years) whereas middle-aged adults

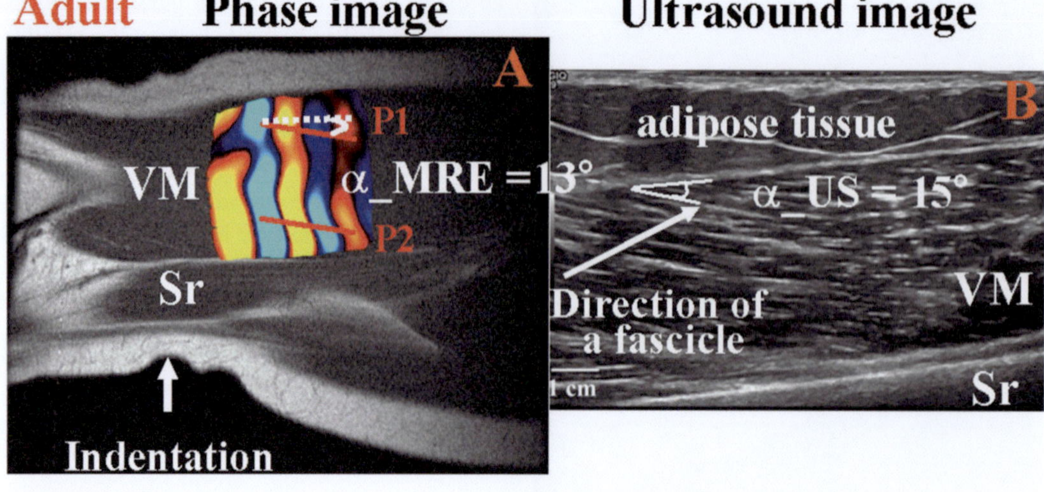

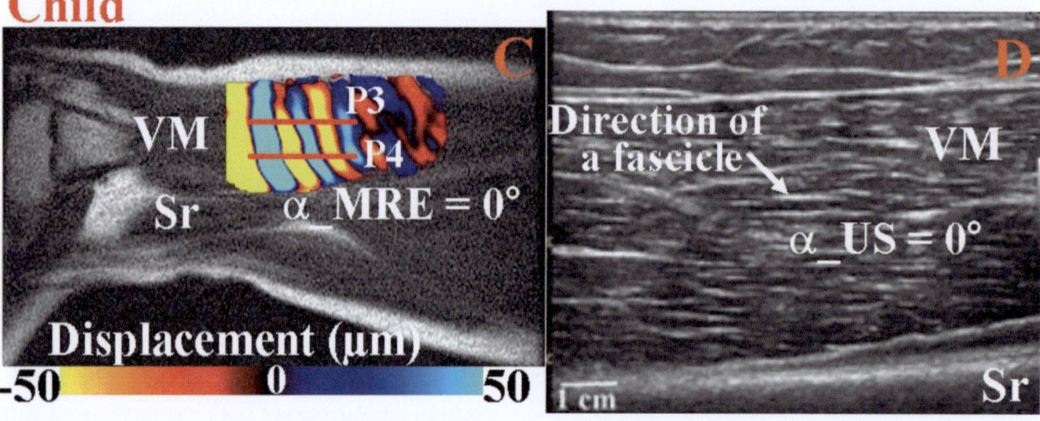

Fig. 7.3 Propagation of shear wave and fascicle angles obtained with MRE (**a–c**) and ultrasound (**b–d**) techniques for adult and child muscle at rest. Profiles (P) were placed in the upper part (P1, P3) of VM and close to the aponeurosis (P2, P4) membrane

(52-59 years) and children (8-12 years) have already reached their maximum recruitment (maximal stiffness) at 10 % of MVC.

After 50 years of age, structural changes, such as increase of collagen cross-link and percentage of slow fibers, force the muscle into developing a high stiffness to perform 10 % of MVC. Then, at 20 % of MVC the slight decrease of muscle stiffness may be also due to the ageing process, i.e. decline of the size and number of muscle fibers with a fatty tissue infiltration. At rest, muscle in children had a significant lower shear modulus compared to those in adults which have similar stiffness [13]. The unorganized muscle structure in the young children (fibers growing and ongoing structural organization) may explain this result which is also confirmed with the high attenuation coefficient (Fig. 7.4b) reflecting a less compact media. However, at 10 % MVC children revealed the highest muscle stiffness showing randomized muscle fibers recruitment.

The attenuation coefficient represents the texture of the medium through which the shear waves are traversing. For instance, subcutaneous adipose muscle tissue had lower shear modulus (μ_fat = 1.73 ± 0.17 kPa) and higher attenuation coefficient (α_fat = 62.71 ± 4.52 m^{-1}) compared to the ones of muscle tissue (μ_muscle = 2.94 ± 0.20 kPa and α_muscle = 50.14 ± 3.08 m^{-1}).

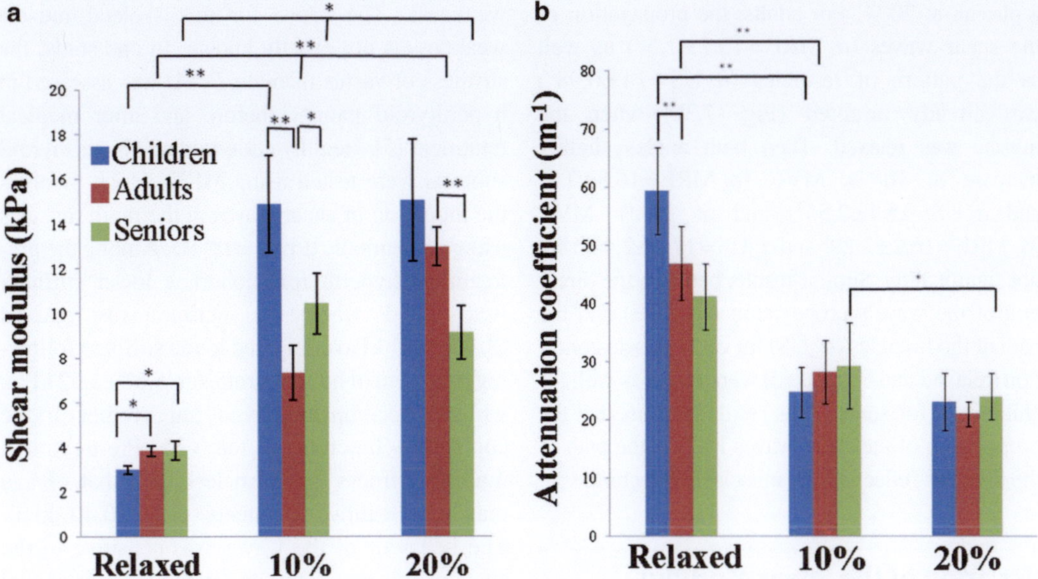

Fig. 7.4 **a**: shear modulus and **b**: attenuation coefficient for children, young, and middle-aged adults when the muscle is relaxed and contracted

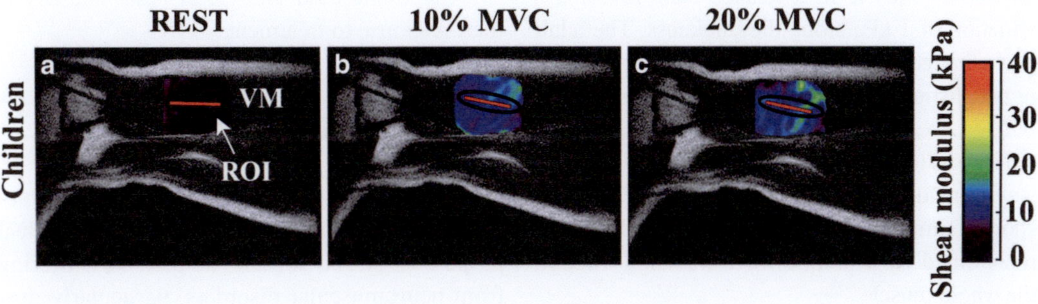

Fig. 7.5 Cartographies of shear modulus at rest (**a**) and contracted conditions (**b** and **c**)

Cartography (Mapping) of Stiffness

Useful information can be obtained by mapping stiffness of muscle in a ROI to see the differences between resting and contracted stages. At rest, the shear modulus map, whatever the age is, shows a homogeneous purple color in the ROI (Fig. 7.5a) placed around the red profile where the shear modulus is initially measured. The comparison of the shear modulus measured inside the ROI and along the red profile was in the same location at rest and in contraction. Figure 7.5b showed a shear modulus map with a diffuse distribution of colors, inside the ROI demonstrating the sensitivity of the shear waves to the muscle internal structure and more specifically to the contraction of the muscle fibers.

Comparison Between the Shear Wave (α_MRE) and the Fascicle Angle (α_US)

The children showed angles (α_MRE and α_US) close to the zero value at rest revealing a muscle structure composed of parallel fascicles. At 10 % MVC, both angles significantly increase (α_MRE = 11.6±3.14° and α_US = 10.6±2.27°), indicating a parallel-oriented pattern of fascicles followed by a plateau at 20 %. For adults, the propagation of the shear waves ($\alpha\_MRE = 13.7 \pm 1.5°$) as well as the pattern of fascicles ($\alpha\_US = 14 \pm 0.98°$) was already oriented (Fig. 7.3b) when the muscle was relaxed. Then both angles slightly increase at 10 % MVC ($\alpha\_MRE = 16.8 \pm 2.6°$ and $\alpha\_US = 15.4 \pm 2.54°$) and at 20 % MVC ($\alpha\_MRE = 16.2 \pm 2.29°$ and $\alpha\_US = 17.2 \pm 2.44°$) but not significantly. Similar results between the direction of the wave's propagation (α_MRE) and the path of the fascicles (α_US) for each muscle condition (relaxed and contracted) were found as well for children as for adults. This result suggests that the propagation of the shear waves follows the path of the fascicles reflecting the muscle architecture.

Variation of the Shear Modulus Inside the Muscle

The measurements of the shear modulus (μ) inside the adults' muscle in a relaxed state revealed a variation of 1 kPa along the thickness. The adult phase image (Fig. 7.3a) showed the propagation of two distinct shear waves in the upper and lower part of the muscle leading to different shear stiffness. However, the propagation of the shear waves inside the muscle of children (Fig. 7.3c) showed similar stiffness at the same localizations, represented by a unique wave traveling through the whole muscle.

The comparison of stiffness between the muscle and subcutaneous adipose tissue showed significant difference between the tissues, with a lower stiffness (1.73 ± 0.17 kPa) for the subcutaneous adipose tissue compared to the one of muscle tissue (2.94 ± 0.20 kPa) and this result was independent of Body Mass Index of the subjects.

Clinical Application of MRE of Muscle

Evaluation of Thyroid Associated Myopathy with MRE

Graves' disease is a common cause of hyperthyroidism which leads to variety of clinical symptoms including reversible skeletal muscles weakness. The cause for this skeletal muscle weakness is not exactly known. In one study, the stiffness of vastus medialis (VM) was assessed in hyperthyroid patients before and after medical treatment [5]. Healthy euthyroid and hyperthyroid subjects were tested using MRE, which involves the induction of shear waves in the thigh muscles using a pneumatic driver at 90 Hz. Among the pre-treatment hyperthyroid cohort, a lower stiffness was found when the muscle was relaxed (2.11 ± 0.61 kPa) compared to the stiffness following treatment of hyperthyroidism (5.52 ± 1.52 kPa), which is accompanied by an improvement in the contractile function of the VM. Pre-treatment muscle stiffness was also less than that of age matched healthy volunteers (4.56 ± 0.40 kPa). The behavior of the waves was sensitive to the stage of this myopathy and to the amount of free thyroxine levels in the serum (FT4). This study shows that MRE provides a non-invasive new tool to evaluate the pathophysiology of thyroid associated and other muscle diseases and assess their response to treatment.

Future Directions

MRE could be a potential clinical tool to characterize the mechanical and the morphological properties of the muscle in patients suffering from neuromuscular disorders, particularly during therapeutic trials. The studies performed on skeletal muscle through the use of MRE technique have demonstrated its capability to characterize the muscle contractile properties in function of age reflected by a mapping of the shear modulus in relation to the level of contraction. Thus, useful information about the spatial distribution of the contracted muscle area related to the physiological activity could be used.

In a study by Chen et al. [14] MRE of the upper trapezius muscle affected by myofascial pain was evaluated in order to visualize and to localize areas of higher stiffness when the muscle was relaxed. Muscle shear modulus measured at different locations revealed that adult muscle tissue is not a homogeneous media while children revealed a homogeneous media due to the ongoing growing process. Stiffness variation found for adult's muscle may be due to differences in fiber tension, which seems to be tenser close to the aponeurosis. This information is of great importance for the follow-up of patients involving a stiffness measurement always in the same muscle area.

Beyond the measured mechanical parameter, MRE can also provide architectural muscle parameter (fascicle angle) and can detect different physiological media. Indeed, stiffness of subcutaneous adipose tissue was found to be significantly lower than the one measured in the belly muscle. This information is of great interest for future MRE experiments performed on patients with myopathy such as Muscular Duchenne Dystrophy since muscle fibers are gradually replaced with adipose tissue. Moreover, the strong correlation between the orientation of the muscle fascicles and the direction of the shear wave propagation attest the assumption that the wave propagation reflects the muscle geometry.

Localization of functional compartments within muscle groups and muscle motion can be performed using vibration MR imaging with MRE [15] and this may be complementary to electromyography for studying normal and abnormal hand biomechanics and before surgical interventions such as tendon transfers or nerve grafting for delineating functional and non-functional muscles for effective interventions.

The association of MRE and ultrasound techniques has demonstrated that the shear modulus is correlated to the muscle architecture. All of these additional data may be correlated to the mechanical properties, which can be used by clinicians to quantify the effects of pathology and treatment, to track the progresses in rehabilitation and to improve treatment methods.

References

1. Bensamoun SF, Ringleb SI, Littrell L, Chen Q, Brennan M, Ehman RL, An KN. Determination of thigh muscle stiffness using magnetic resonance elastography. J Magn Reson Imaging. 2006;22:242–7.
2. Bensamoun SF, Glaser KJ, Ringleb SI, Chen Q, Ehman RL, An KN. Rapid magnetic resonance elastography of skeletal muscle using one dimensional projection. J Magn Reson Imaging. 2008;27:1083–8.
3. Dresner MA, Rose GH, Rossman PJ. Magnetic resonance elastography of skeletal muscle. J Magn Reson Imaging. 2001;13:269–76.
4. Chen Q, Basford J, An KN. Ability of magnetic resonance elastography to assess taut bands. Clin Biomech. 2008;23:623–9.
5. Bensamoun SF, Ringleb SI, Chen Q, Ehman RL, An KN, Brennan M. Thigh muscle stiffness assessed with magnetic resonance elastography in hyperthyroid patients before and after medical treatment. J Magn Reson Imaging. 2007;26:708–13.
6. Ringleb SI, Bensamoun SF, Chen Q, Manduca A, Ehman RL, An KN. Applications of magnetic resonance elastography to healthy and pathologic skeletal muscle. J Magn Reson Imaging. 2007;25(2):301–9. Invited Review.
7. Jenkyn TR, Ehman RL, An KN. Noninvasive muscle tension measurement using the novel technique of magnetic resonance elastography (MRE). J Biomech. 2003;36:1917–21.
8. Basford JR, Jenkyn TR, An KN, Ehman RL, Heers G, Kaufman KR. Evaluation of healthy and diseased muscle with magnetic resonance elastography. Arch Phys Med Rehabil. 2002;83:1530–6.
9. McCullough MB, Domire ZJ, Reed AM, Amin S, Ytterberg SR, Chen Q, An KN. Evaluation of muscles affected by myositis using magnetic resonance elastography. Muscle Nerve. 2011;43:585–90.
10. Debernard L, Leclerc GE, Robert L, Charleux F, Bensamoun SF. In vivo characterization of the muscle viscoelasticity using mutifrequency MR elastography (MMRE). J Musculoskelet Res. 2013;16:1350008-1–10.
11. Manduca A, Oliphant TE, Dresner MA, Mahowald JL, Kruse SA, Amromin E, Felmlee JP, Greenleaf JF, Ehman RL. Magnetic resonance elastography: noninvasive mapping of tissue elasticity. Med Image Anal. 2001;5:237–54.
12. Debernard L, Robert L, Charleux F, Bensamoun S. Characterization of muscle architecture in children and adults using magnetic resonance elastography and ultrasound techniques. J Biomech. 2011;44:397–401.
13. Debernard L, Robert L, Charleux F, Bensamoun S. Analysis of thigh muscle stiffness from childhood to adulthood using magnetic resonance elastography (MRE) technique. Clin Biomech. 2011;26:836–40.
14. Chen Q, Bensamoun SF, Basford JR, Thompson JM, An KN. Quantification of muscle pathophysiology with magnetic resonance elastography. Arch Phys Med Rehabil. 2007;88(12):1658–61.
15. Mariappan YK, Manduca A, Glaser KJ, Chen J, Amrami KK, Ehman RL. Vibration imaging for localization of functional compartments of the extrinsic flexor muscles of the hand. J Magn Reson Imaging. 2010;31:1395–401.

Magnetic Resonance Elastography of the Brain

8

John Huston, III

Introduction

The human brain is perhaps the most imaged organ in the body. The brain is affected by many disorders including infections, stroke, tumors, and neurodegenerative disorders. Most of these cerebral pathologies can be detected and characterized by the existing state of the art imaging methods; however some diseases, such as neurodegenerative disorders like Alzheimer's disease, still pose a challenge. Subtle structural alterations in the brain parenchyma occur in neurodegenerative disorders that often go undetected with conventional imaging techniques. Normal pressure hydrocephalus (NPH) is thought to be caused by changes in mechanical properties of the brain [1]. Assessing mechanical properties of brain may therefore be useful to detect the structural changes giving insights to diffuse pathological processes and a method to characterize focal lesions. Palpation of the brain has been limited to neurosurgeons during open surgery and is qualitative. With the introduction of the MR Elastography (MRE) technique it is now possible to non-invasively evaluate mechanical properties of the brain parenchyma and intracranial tumors.

Quantitative measurement of viscoelastic properties of brain seated within the bony cranium is challenging and magnetic resonance based elastography has recently been successful in achieving this. Currently, several methods for generating and then imaging the propagating mechanical waves in the brain with MRE exist [2–5]. The MRE technique has been useful in demonstrating different viscoelastic properties of the brain in Alzheimer's disease [6], multiple sclerosis [7], and progressive supranuclear palsy [8]. The technique of brain MRE continues to evolve with modifications to reduce imaging time and improve resolution setting the stage for more clinical applications.

Performing Brain MRE

Several different methods for performing MRE of the brain have been described. The shear waves have been introduced into the brain using a electromechanical actuator [9], a thermoplastic bite block attached to actuator [10], a custom made cradle connected to a remote vibration generator "head-rocker unit" [4], and a soft pillow driver [6]. A novel method of MRE of brain using MR table vibrations resulting from fast switching of the gradient coils (TREMR) is possible [11]. MRE using intrinsic motion caused by pulsatile flow of blood/blood pressure termed as auto-elastography has also been described [11].

J. Huston, III, M.D. (✉)
Department of Radiology, Mayo Clinic,
200 First Street SW, Rochester, MN 55905, USA
e-mail: jhuston@mayo.edu

S.K. Venkatesh and R.L. Ehman (eds.), *Magnetic Resonance Elastography*,
DOI 10.1007/978-1-4939-1575-0_8, © Springer Science+Business Media New York 2014

We will describe here the MRE technique using soft pillow driver currently used in both research and clinical studies at our institute.

Driver

MRE of the brain is possible on both 1.5T and 3T MR scanners. The hardware required for the MRE is a soft pillow like vibration source using a pneumatic actuator. The active component of the actuator, located outside of the scan room, is comprised of a waveform generator, an amplifier, and an acoustic speaker. The passive pillow-like component consists of a soft, inelastic, fabric cover over a porous, springy, mesh measuring 15×9×1.5 cm. The soft vibration source is placed under the subject's head within an 8-channel receive-only head coil (Fig. 8.1). The active and passive driver components are connected by a 24-foot long, 0.75-inch diameter flexible tube from the active driver terminating in a 0.5-inch diameter, 1.5-foot long tube integrated into the passive driver.

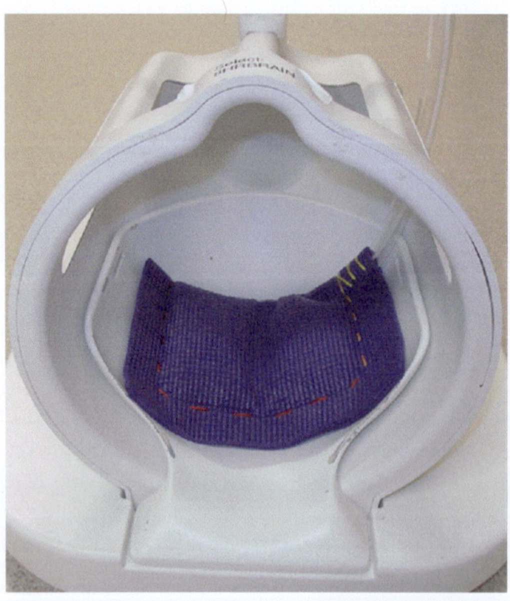

Fig. 8.1 Soft pillow driver used to introduce shear waves into the brain positioned within a standard eight channel head coil

Pulse Sequence

Brain MRE data is typically acquired using a modified single-shot, spin-echo EPI pulse sequence. Shear waves of 60 Hz are introduced using the pneumatic active driver (located outside of the scanner room) and a soft, pillow-like passive driver placed under the subject's head as previously described [6]. The resulting motion is imaged with the following parameters: TR/TE=3,600/62 ms; field of view (FOV)=24 cm; BW=±250 kHz; 72×72 imaging matrix reconstructed to 80×80; frequency encoding in the right-left direction; 3x ASSET (SENSE) acceleration; 48 contiguous 3 mm thick axial slices; one 4 G/cm, 18.2 ms, zeroth- and first-order moment nulled motion encoding gradient on each side of the refocusing RF pulse synchronized to the motion; motion encoding in the positive and negative x, y and z directions; and eight phase offsets sampled over one period of the 60 Hz motion. The resulting images have 3 mm isotropic resolution and are acquired in less than 7 min.

Meningioma

The application of brain MRE to focal tumors has been proposed [12] and has already been found useful for evaluating meningiomas prior to resection [13]. Most meningiomas are characterized on MRI imaging by isointense T1 and T2 signal with uniform gadolinium contrast enhancement. However, despite their frequent uniform appearance, their viscoelastic properties range widely from very soft and removable by suction to very stiff requiring extensive dissection including the use of ultrasonic aspirators. While it is known that extremely soft meningiomas are seen on MRI as increased T2 signal with decreased T1 signal, and that extremely stiff meningiomas present with decreased T2 signal, the physical characteristics of the vast majority of meningiomas are unknown prior to surgery [14]. This fact impacts clinical care because extremely stiff meningiomas, especially those located at the skull base, are associated with increased surgical risk due to the

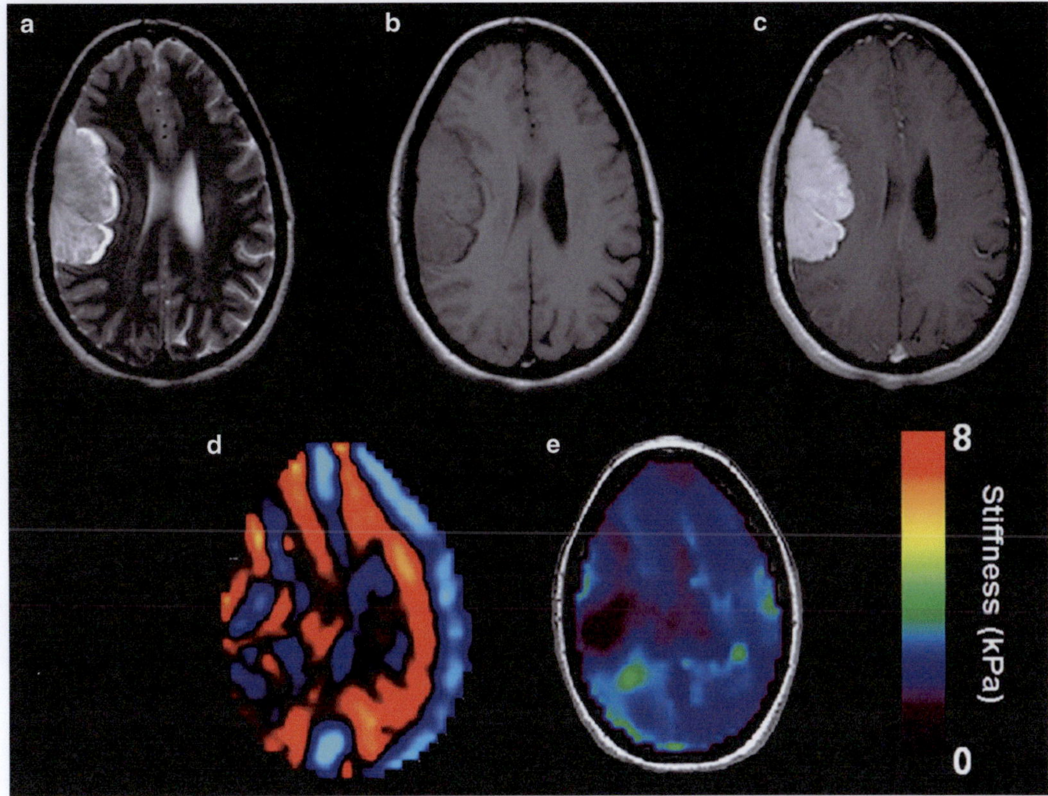

Fig. 8.2 Approximate 6×3.5 cm dural based extraaxial tumor with mildly increased T2 signal (**a**) and isointense T1 signal (**b**). Relatively uniform contrast enhancement following the administration of gadolinium contrast (**c**). Wave image (**d**) and the elastogram (**e**) show the meningioma is softer than surrounding brain tissue. This was confirmed at surgery where the lesion was easily resected with suction

potential for damage to adjacent and encompassed arteries, veins, and nerves. Various techniques have been attempted to preoperatively identify stiffness including standard MRI [14, 15], MR spectroscopy [16], MR diffusion imaging [17], and CT [18].

MRE has been shown to be a good preoperative assessment tool for meningioma properties, improving both risk assessment and patient management [13]. A series of meningiomas were evaluated using a five grade scale, ranging from soft and 100 % removable with suction to uniformly hard, requiring that most of the tumor be removed with ultrasonic aspiration (Fig. 8.2). Meningioma stiffness measured by MRE compared to the surgeons' qualitative assessment of tumor stiffness determined at the time of surgery showed stiffness significantly correlated with the qualitative assessment at surgery ($p = 0.023$) [13].

The introduction of high resolution MRE with 3 mm isotropic spatial resolution has allowed the identification of internal heterogeneity within meningiomas (Fig. 8.3). As meningiomas enlarge the internal tumor components may develop variable regions of stiffness. Often calcification either on the tumor surface or internally leads to considerable stiffening. Also, the density of the fibrotic components of tumor without frank calcification may lead to very stiff tissue. Internal necrosis due to ischemic or hemorrhagic change may result in regional softening (Fig. 8.3). To provide the surgeon with a preoperative map of the variable internal tissue characteristics may lead to altered surgical plans with a safer potentially more complete resection of the lesion.

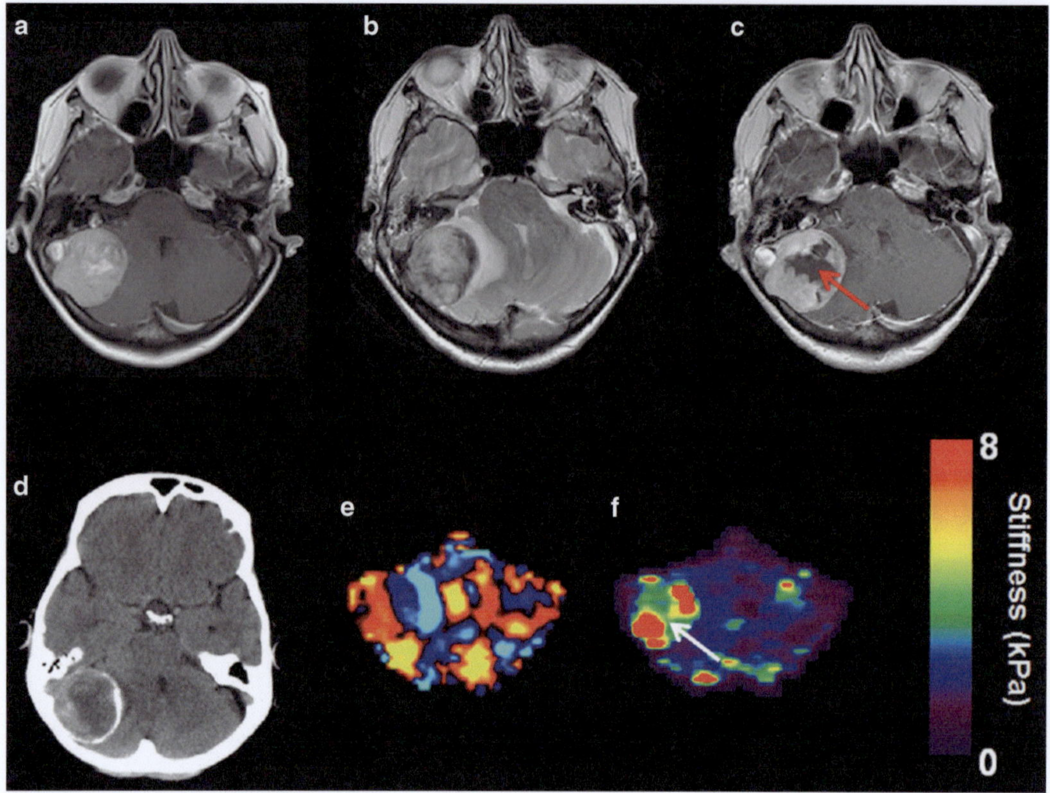

Fig. 8.3 Example of an approximate 4 cm right posterior fossa meningioma with varied internal viscoelastic characteristics. MRI T1 weighted post gadolinium image from 2011 demonstrates uniform enhancement throughout the lesion (**a**). T2 weighted image from a followup 2014 exam shows the interval development of surrounding vasogenic edema and increased mass effect (**b**). T1 weighted post gadolinium image shows central loss of contrast enhancement that corresponded to histological changes of hemorrhagic necrosis (*arrow* **c**). CT demonstrated peripheral calcification (**d**). Wave image shows the shear wavelength is longer in this stiff meningioma (**e**). Elastogram shows heterogeneity corresponding to the very stiff peripheral tumor and soft central necrosis (*arrow* **f**)

Glioma

Gliomas are the most frequent type of primary tumors arising in the brain. A biopsy is required for accurate pathologic characterization of the tumor grade and is based on the World Health Organization grading system [19]. The most common type of gliomas are astrocytomas which are classified as pilocytic astrocytoma (grade I), low-grade astrocytoma (grade II), anaplastic astrocytoma (grade III), and glioblastoma multiforme (grade IV). While standard imaging findings can be suggestive such as the presence of contrast enhancement and necrosis, no non-invasive technique exists for the preoperative classification of glioma tumor grade [20]. The potential exists that the addition of stiffness information may allow improved preoperative characterization of glioma subtypes (Fig. 8.4).

The possibility of utilizing MRE as a biomarker to determine cancer therapy response has already been demonstrated in non-Hodgkin's lymphoma and colon cancer [21, 22]. Following chemotherapy, MRE measured a significant decrease in tumor stiffness before a clinically significant change in tumor volume, suggesting MRE may offer an earlier indication of treatment response compared with current clinical methods. Changes in the mechanical properties of the extracellular matrix

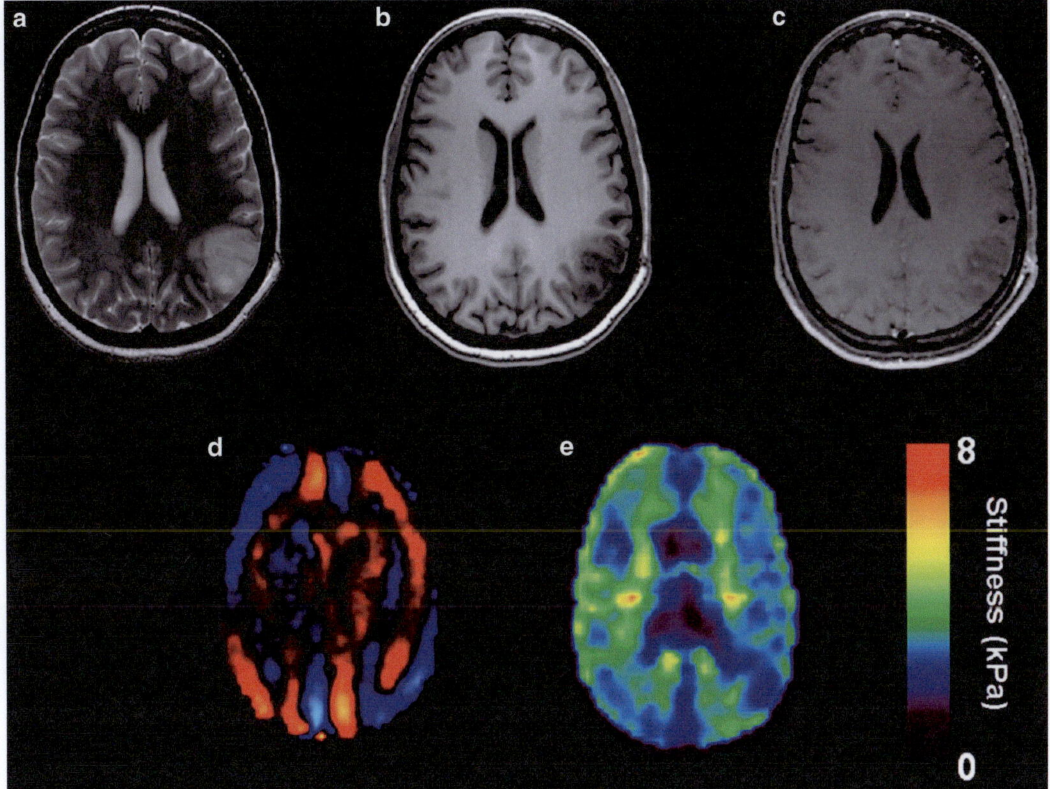

Fig. 8.4 Example of an oligodendroglioma in a 34 year old female discovered following a single seizure. T2 weighted image demonstrates an infiltrative lesion with involvement of both the white matter and overlying cortex with enlargement of the gyri typical of oligodendroglioma (**a**). Low T1 signal is seen before the administration of gadolinium (**b**) with no definite enhancement demonstrated (**c**). Wave image (**d**) and elastogram (**e**) shows the tumor to be softer than surrounding normal brain tissue

have been shown to influence cancer progression [23], where stiffening of the matrix promotes the spread of malignant cells. MRE is capable of quantifying change in the tissue mechanical properties, and may be able to identify increased brain tissue stiffness surrounding the tumor that may indicate the spread of disease. A noninvasive assessment of tumor spread and a prediction of recurrence would have a significant impact on patient care.

Pituitary Adenoma

Pituitary adenomas are a common tumor of pituitary gland that may result in abnormally increased hormone production. Some adenomas are not associated with endocrine changes and as a result may reach large size before indirect symptoms lead to their discovery. When possible transsphenoidal is the preferred surgerical approach for resection of the tumor [24]. Surgical risk increases with increased size, lobulation, invasion of the adjacent cavernous sinuses, and increased stiffness of tumor [25]. Soft lesions are often removed with suction through a transsphenoidal approach. However, stiff lesions can be very difficult to remove in large part due to the limited surgical operating field of view and therefore an open craniotomy is preferred. MRE offers a preoperative evaluation of tumor stiffness and improved surgical planning for the safest resection technique (Fig. 8.5).

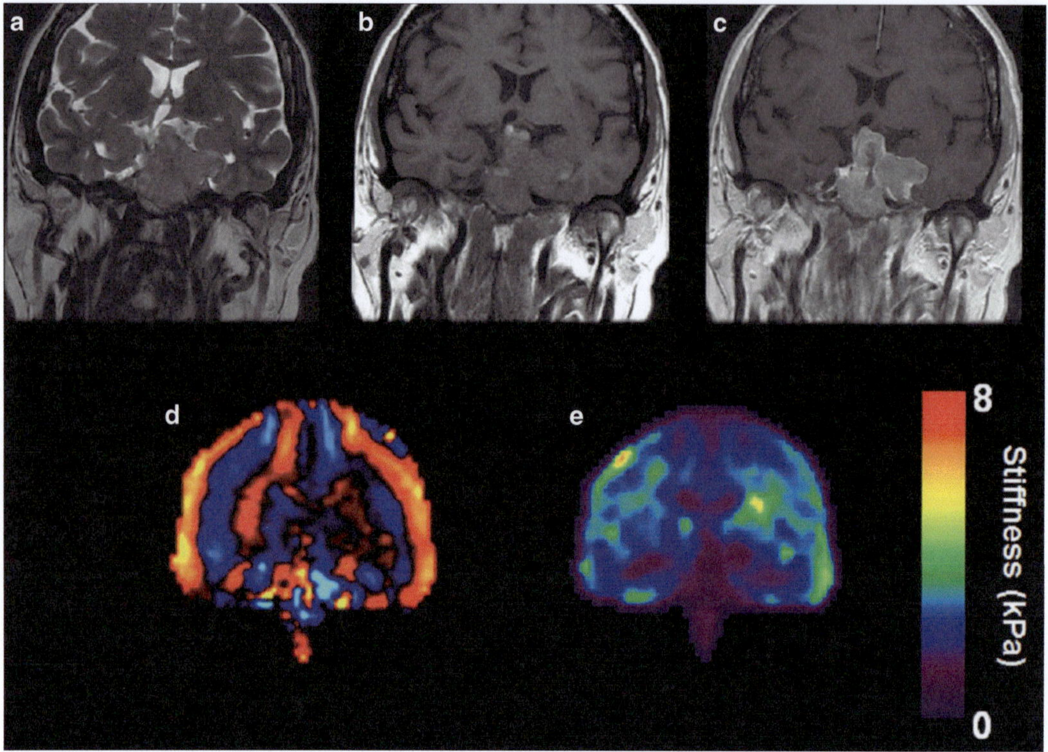

Fig. 8.5 Very large sellar mass invading the left cavernous sinus and clivus extending into and filling the sphenoid sinus which was demonstrated to represent a null cell pituitary adenoma. Coronal imaging shows the tumor to be essentially isointense to brain tissue on T2 weighted (**a**) and T1 weighed (**b**) sequences. Mildly irregular enhancement is seen following the administration of gadolinium (**c**). Wave image (**d**) and elastogram (**e**) shows the tumor to be softer than brain tissue which was confirmed when the tumor was removed

Demyelination

MRI plays a prominent role in the diagnosis and management of demyelinating diseases such as multiple sclerosis (MS). MS is an immune mediated disease in which monocytes and T-cells induce an inflammatory response that causes both demyelination and degeneration of axons and gliosis. As the body's immune cells damage the myelin, axons lose the ability to conduct electrical signals resulting in functional impairment. Recent studies demonstrate that the brain tissue involved in MS is not confined to the white matter, and there is a growing understanding of the role of cortical involvement [26]. MS is thus considered a diffuse disease of the central nervous system. Neuroimaging has come to be a standard tool for diagnosing MS. Gadolinium enhancement, the result of a breakdown of the blood–brain-barrier, is one of the earliest imaging findings and typically indicates active disease. The McDonald criterion, introduced in 2001 for the diagnosis of MS, gave MRI a major role [27]. MRI also has been used as a surrogate for clinical trials of new treatments. However, standard MRI findings are not always strongly correlated with disease severity or predictive of future relapse [28], a situation referred to as the "clinicoradiologic paradox." To overcome this conundrum, a technique such as MRE that evaluates the structural integrity of brain tissue may have a role by characterizing tissue alterations not detected by standard imaging.

Wuerfel reported the results of an MRE study in which four different driver frequencies were used in 45 patients with mild relapsing-remitting MS. The study was conducted with a 1.5T scanner with a single-shot spin-echo EPI sequence [7].

They found that at 62.5 Hz, there was a 13 % decrease in the viscoelasticity in MS patients compared to matched healthy controls. This decreased stiffness was independent of standard MRI findings such as volume of T2 signal abnormality or enhancement. A follow-up study was conducted on 23 patients, focusing on chronic-progressive MS and utilizing similar techniques [29]. In these chronic MS patients the reduction in stiffness was 20 %. This decreased stiffness was independent of age, brain parenchymal fraction, and total brain volume in a multivariate analysis. The larger reduction of stiffness in chronic-progressive MS may well reflect the mechanical disruption of the underlying brain parenchyma due to extended exposure to the inflammatory effects of the disease.

Hydrocephalus and NPH

Hydrocephalus occurs with the expansion of the ventricular system resulting from increased intra-ventricular pressure due to obstruction of the ventricular outlet, obstruction of CSF resorption by the arachnoid villi or rarely overproduction of CSF. Pattison investigated an experimental model of hydrocephalus in cats induced by intrathecal injection of kaolin [30]. Twelve cats were injected into the cisterna magna with typical dose of kaolin 10 mg in 0.25 ml saline while six normal controls with injection of normal saline. A representative cat had global shear modulus at baseline of 5.5 increased to 9.4 kPa. T2 weighted imaging demonstrated an increase in ventricular size from 0.2 to 1.49 cm^3 ($p < .0001$) in the kaolin injected cats while the controls experienced a non-significant increase in ventricular volume.

NPH was described by Adams et al. [31]. The classic symptom triad includes cognitive impairment, urinary incontinence and gait disturbance, but the diagnosis can be challenging. The incidence of clinically suspected NPH is estimated at 3.74/100,000/year for the total population and at 15.20/100,000/year for the population older than 50 years [32]. Cross sectional imaging demonstrates enlarged ventricles out of proportion to the size of the sulci but increased intracranial pressure is not found by diagnostic lumbar puncture is performed. However, clinical improvement results from therapeutic drainage of CSF either for diagnostic purposes (such as a 50 cc diagnostic tap) or with placement of a ventricular shunt tube. The etiology of NPH remains elusive.

Streitberger et al. studied changes in brain viscoelastic properties resulting from NPH in 20 patients with a mean age of 69 and in 25 age matched normal controls (33_ENREF_33). MRE exams were performed using a 1.5T system with a head cradle to induce shear waves. The complex shear modulus calculated at 62.5 Hz was 2.42 kPa for the NPH subjects compared to 2.66 kPa for the normal controls with a decrease in elasticity for the NPH subjects corresponding to an increase in the compliance of the brain tissue (Fig. 8.6).

Alzheimer's Disease

The symptoms of Alzheimer's disease (AD) include progressive decline in memory, cognitive abilities, language, and motor skills. AD pathology includes the accumulation of extracellular amyloid plaques, intracellular neurofibrillary tangles, and neurodegeneration. These abnormal proteins are many magnitudes greater in stiffness than brain parenchyma, and it was assumed that the impact of their accumulation would increase brain stiffness in AD patients.

To assess the impact of AD on the viscoelastic properties of the brain, we conducted a study of 28 subjects, including 7 with probable AD, 14 age and gender matched cognitively normal controls without elevated levels of amyloid and 7 age and gender matched cognitively normal controls with elevated levels of abnormal amyloid. The MRE results showed significantly less brain stiffness in AD subjects compared to both groups of controls (Fig. 8.7). The median stiffness of the cognitively normal controls without amyloid was 2.37 kPa; the median stiffness for cognitively normal group with abnormal amyloid was 2.32 kPa; and the median stiffness for the AD group was 2.20 kPa. The cognitively normal groups did not differ from

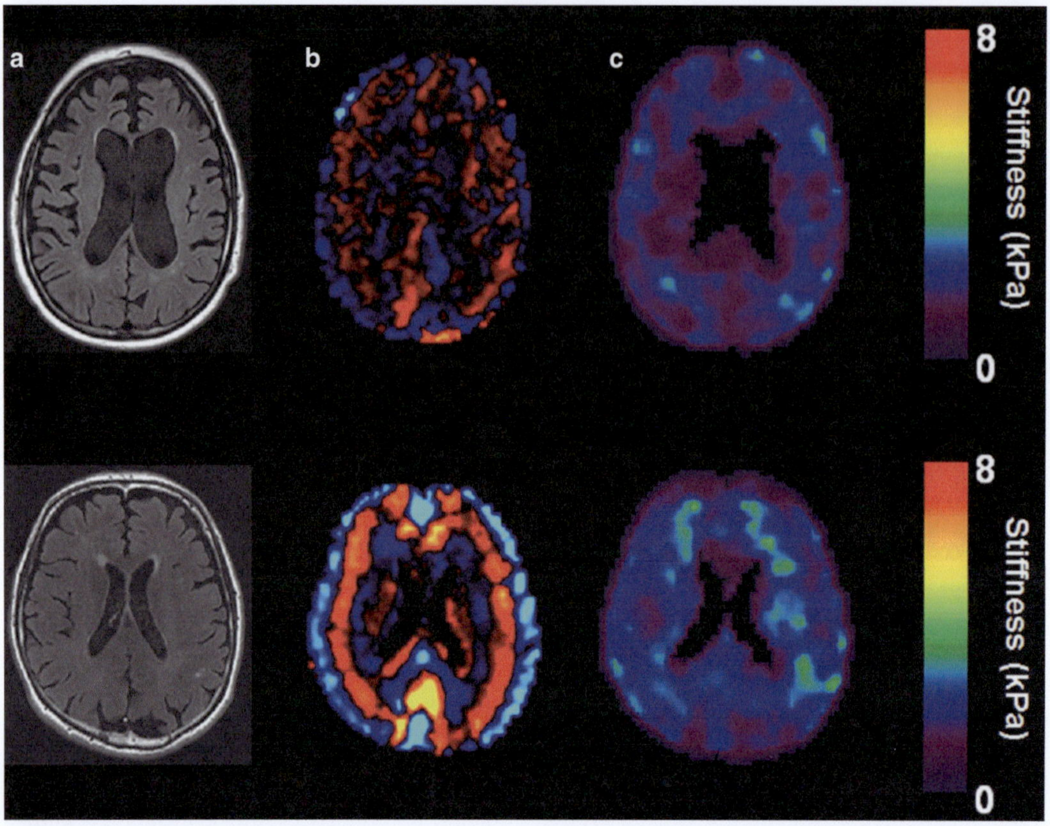

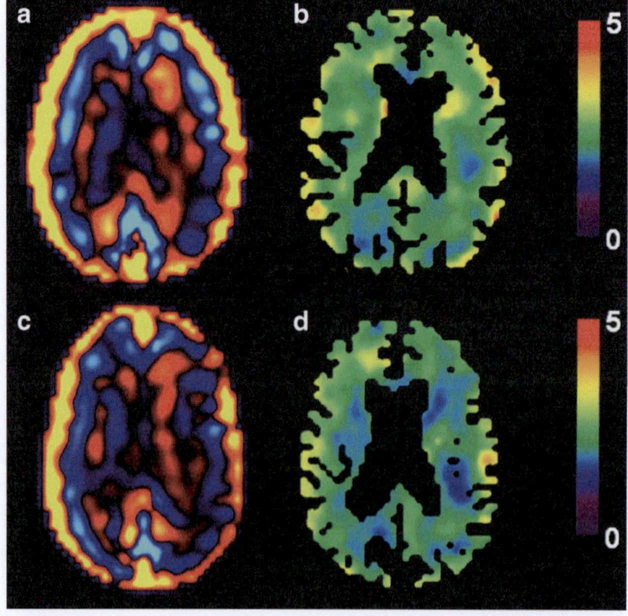

Fig. 8.6 Normal pressure hydrocephalus in a 75 year old male who had symptoms and imaging consistent with NPH including enlarged ventricles out of proportion to the size of the sulci and a CSF flow study with hyperdynamic CSF flow (*upper set* of images) compared with an age matched normal male (*lower set* of images). T2 FLAIR images demonstrate the enlarged ventricles of the NPH patient compared to the control (**a**). Wave images (**b**) and elastograms (**c**) show the brain parenchyma is softer in the NPH patient

Fig. 8.7 Eight-nine year old with Alzheimer's (*lower set of images*) compared with a 93 year old cognitively normal control (*upper set of images*). Wave image (**c**) and elastogram (**d**) demonstrate that the brain parenchyma is softer in the Alzheimer's disease subject compared to the wave image (**a**) and elastogram (**b**) of the normal control

each other ($p=0.85$), however the AD group had a significantly lower stiffness than the control group without amyloid ($p=0.0015$) and the group with amyloid ($p=0.026$).

Conclusion

Brain MRE has been technically challenging to develop but already has demonstrated clinical utility in both diffuse and focal diseases. Continued development of the technique will expand our understanding of neurological process and disease as well as offer improved patient care.

References

1. Bradley Jr WG, Whittemore AR, Watanabe AS, Davis SJ, Teresi LM, Homyak M. Association of deep white matter infarction with chronic communicating hydrocephalus: implications regarding the possible origin of normal-pressure hydrocephalus. AJNR Am J Neuroradiol. 1991;12(1):31–9.
2. Kruse SA, Rose GH, Glaser KJ, Manduca A, Felmlee JP, Jack Jr CR, et al. Magnetic resonance elastography of the brain. Neuroimage. 2008;39(1):231–7.
3. Green MA, Bilston LE, Sinkus R. In vivo brain viscoelastic properties measured by magnetic resonance elastography. NMR Biomed. 2008;21(7):755–64 [Research Support, Non-U.S. Gov't].
4. Sack I, Beierbach B, Hamhaber U, Klatt D, Braun J. Non-invasive measurement of brain viscoelasticity using magnetic resonance elastography. NMR Biomed. 2008;21(3):265–71 [Research Support, Non-U.S. Gov't].
5. Hamhaber U, Sack I, Papazoglou S, Rump J, Klatt D, Braun J. Three-dimensional analysis of shear wave propagation observed by in vivo magnetic resonance elastography of the brain. Acta Biomater. 2007;3(1): 127–37 [Research Support, Non-U.S. Gov't].
6. Murphy MC, Huston 3rd J, Jack Jr CR, Glaser KJ, Manduca A, Felmlee JP, et al. Decreased brain stiffness in Alzheimer's disease determined by magnetic resonance elastography. J Magn Reson Imaging. 2011;34(3):494–8 [Research Support, N.I.H., Extramural].
7. Wuerfel J, Paul F, Beierbach B, Hamhaber U, Klatt D, Papazoglou S, et al. MR-elastography reveals degradation of tissue integrity in multiple sclerosis. Neuroimage. 2010;49(3):2520–5 [Research Support, Non-U.S. Gov't].
8. Lipp A, Trbojevic R, Paul F, Fehlner A, Hirsch S, Scheel M, et al. Cerebral magnetic resonance elastography in supranuclear palsy and idiopathic Parkinson's disease. Neuroimage Clin. 2013;3:381–7.
9. Kruse SA, Dresner MA, Rossman P, Femlee JP, Jack CR, Jr., Ehman RL, editors. Palpation of the brain using magnetic resonance elastography. International Society for Magnetic Resonance in Medicine; 22–28 May 1999; Philadelphia, PA.
10. Kruse SA, Ehman RL, editors. 2D approximation of 3D wave propagation in MR elastography of the brain. In: Proceedings of the 11th international society for magnetic resonance in medicine; 10–16 July 2003; Toronto, Ontario, Canada.
11. Gallichan D, Robson MD, Bartsch A, Miller KL. TREMR: table-resonance elastography with MR. Magn Reson Med. 2009;62(3):815–21 [Research Support, Non-U.S. Gov't].
12. Xu L, Lin Y, Han JC, Xi ZN, Shen H, Gao PY. Magnetic resonance elastography of brain tumors: preliminary results. Acta Radiol. 2007;48(3):327–30 [Research Support, Non-U.S. Gov't].
13. Murphy MC, Huston 3rd J, Glaser KJ, Manduca A, Meyer FB, Lanzino G, et al. Preoperative assessment of meningioma stiffness using magnetic resonance elastography. J Neurosurg. 2013;118(3):643–8 [Evaluation Studies Research Support, N.I.H., Extramural].
14. Hoover JM, Morris JM, Meyer FB. Use of preoperative magnetic resonance imaging T1 and T2 sequences to determine intraoperative meningioma consistency. Surg Neurol Int. 2011;2:142.
15. Yamaguchi N, Kawase T, Sagoh M, Ohira T, Shiga H, Toya S. Prediction of consistency of meningiomas with preoperative magnetic resonance imaging. Surg Neurol. 1997;48(6):579–83.
16. Chernov MF, Kasuya H, Nakaya K, Kato K, Ono Y, Yoshida S, et al. (1)H-MRS of intracranial meningiomas: what it can add to known clinical and MRI predictors of the histopathological and biological characteristics of the tumor? Clin Neurol Neurosurg. 2011;113(3):202–12 [Research Support, Non-U.S. Gov't].
17. Kashimura H, Inoue T, Ogasawara K, Arai H, Otawara Y, Kanbara Y, et al. Prediction of meningioma consistency using fractional anisotropy value measured by magnetic resonance imaging. J Neurosurg. 2007; 107(4):784–7 [Clinical Trial Research Support, Non-U.S. Gov't].
18. Kendall B, Pullicino P. Comparison of consistency of meningiomas and CT appearances. Neuroradiology. 1979;18(4):173–6.
19. Kleihues P, Sobin LH. World Health Organization classification of tumors. Cancer. 2000;88(12):2887.
20. Chawalparit O, Sangruchi T, Witthiwej T, Sathornsumetee S, Tritrakarn S, Piyapittayanan S, et al. Diagnostic performance of advanced MRI in differentiating high-grade from low-grade gliomas in a setting of routine service. J Med Assoc Thai. 2013;96(10):1365–73 [Research Support, Non-U.S. Gov't].
21. Juge L, Doan BT, Seguin J, Albuquerque M, Larrat B, Mignet N, et al. Colon tumor growth and antivascular treatment in mice: complementary assessment with MR elastography and diffusion-weighted MR imaging. Radiology. 2012;264(2):436–44 [Research Support, Non-U.S. Gov't].

22. Pepin KM, Chen J, Glaser KJ, Mariappan YK, Reuland B, Ziesmer S, et al. MR elastography derived shear stiffness—a new imaging biomarker for the assessment of early tumor response to chemotherapy. Magn Reson Med. 2014;71(5):1834–40.

23. Lu P, Weaver VM, Werb Z. The extracellular matrix: a dynamic niche in cancer progression. J Cell Biol. 2012;196(4):395–406 [Research Support, N.I.H., Extramural Research Support, Non-U.S. Gov't Review].

24. Bodhinayake I, Ottenhausen M, Mooney MA, Kesavabhotla K, Christos P, Schwarz JT, et al. Results and risk factors for recurrence following endoscopic endonasal transsphenoidal surgery for pituitary adenoma. Clin Neurol Neurosurg. 2014;119:75–9 [Research Support, N.I.H., Extramural].

25. Halvorsen H, Ramm-Pettersen J, Josefsen R, Ronning P, Reinlie S, Meling T, et al. Surgical complications after transsphenoidal microscopic and endoscopic surgery for pituitary adenoma: a consecutive series of 506 procedures. Acta Neurochir (Wien). 2014; 156(3):441–9.

26. Lucchinetti CF, Popescu BF, Bunyan RF, Moll NM, Roemer SF, Lassmann H, et al. Inflammatory cortical demyelination in early multiple sclerosis. N Engl J Med. 2011;365(23):2188–97 [Research Support, N.I.H., Extramural Research Support, Non-U.S. Gov't].

27. McDonald WI, Compston A, Edan G, Goodkin D, Hartung HP, Lublin FD, et al. Recommended diagnostic criteria for multiple sclerosis: guidelines from the international panel on the diagnosis of multiple sclerosis. Ann Neurol. 2001;50(1):121–7 [Guideline Research Support, Non-U.S. Gov't].

28. Fisniku LK, Brex PA, Altmann DR, Miszkiel KA, Benton CE, Lanyon R, et al. Disability and T2 MRI lesions: a 20-year follow-up of patients with relapse onset of multiple sclerosis. Brain. 2008;131(Pt 3):808–17 [Research Support, Non-U.S. Gov't].

29. Streitberger KJ, Sack I, Krefting D, Pfuller C, Braun J, Paul F, et al. Brain viscoelasticity alteration in chronic-progressive multiple sclerosis. PLoS One. 2012; 7(1):e29888 [Research Support, Non-U.S. Gov't].

30. Pattison AJ, Lollis SS, Perrinez PR, Weaver JB, Paulsen KD, editors. MR elastography of hydrocephalus. Medical Imaging 2009: Biomedical Applications in Molecular, Structural, and Functional Imaging, 72620A; 27 Feb 2009.

31. Adams RD, Fisher CM, Hakim S, Ojemann RG, Sweet WH. Symptomatic occult hydrocephalus with "normal" cerebrospinal-fluid pressure. a treatable syndrome. N Engl J Med. 1965;273:117–26.

32. Klassen BT, Ahlskog JE. Normal pressure hydrocephalus: how often does the diagnosis hold water? Neurology. 2011;77(12):1119–25 [Comparative Study Research Support, N.I.H., Extramural].

33. Streitberger KJ, Wiener E, Hoffmann J, Freimann FB, Klatt D, Braun J, et al. In vivo viscoelastic properties of the brain in normal pressure hydrocephalus. NMR Biomed. 2011;24(4):385–92 [Research Support, Non-U.S. Gov't].

Magnetic Resonance Elastography of the Lungs

Kiaran P. McGee, Yogesh Mariappan,
Rolf D. Hubmayr, Anthony Romano,
Armando Manduca, and Richard L. Ehman

Anatomy, Physiology, and Mechanical Basis for Lung Function, Disease Initiation, and Progression

The respiratory system consists of the nasal and oral cavities, the larynx and trachea, bronchi and lungs. The conical shaped left and right lungs are dissected by fissures that further sub divide the lung into individual lobes (two within the left and three within the right lung). Each lobe consists of both conductive (trachea and bronchi) and functional units (alveoli). Conductive tissues facilitate gas flow into and out of the lung while functional units perform the task of exchanging O_2 and CO_2 into and out of the blood supply respectively. While the lungs perform several functions including maintenance of acid-base balance, phonation, pulmonary defense, and metabolism and the handling of bioactive materials, their primary function is gas exchange [1].

In terms of tissue mechanical properties, the lungs exist in a state of constant pre-stress as a result of the dynamic interplay between transpulmonary pressure, the elastic properties of lung parenchyma, and the surfactant induced surface tension forces. Under normal conditions a dynamic balance between the intrinsic and applied forces is maintained with stress being distributed throughout the connective tissue network of the parenchyma [2]. Disease processes such as interstitial lung disease (ILD) are known to perturb this balance, resulting in changes to both local and global tissue mechanics [3].

Although the link between lung mechanics and disease initiation and progression is generally appreciated, the ability to assess, in particular at the regional level remains limited. Several methods exist to provide various measures of lung function including direct mechanical testing, bronchoscopy with bronchoalveolar lavage, spirometry-based pulmonary function tests, arterial blood gas analysis, and several imaging based methods including conventional radiography, computed tomography (CT), and magnetic resonance (MR) imaging methods. While each of these tests provides important diagnostic information, their common limitation is the inability to quantitate the intrinsic mechanical properties of lung parenchyma in vivo. Within this context,

K.P. McGee, Ph.D. (✉)
United States Naval Research Laboratory,
Washington, DC, USA
e-mail: McGee.Kiaran@mayo.edu

Y. Mariappan, Ph.D.
Philips Healthcare, Bengaluru, Karnataka, India

R.D. Hubmayr, M.D. • A. Romano, Ph.D.
A. Manduca, Ph.D. • R.L. Ehman, M.D.
Department of Radiology, Mayo Clinic College of Medicine, Mayo Clinic, 200 First Street SW, Rochester, MN, USA

S.K. Venkatesh and R.L. Ehman (eds.), *Magnetic Resonance Elastography*,
DOI 10.1007/978-1-4939-1575-0_9, © Springer Science+Business Media New York 2014

Magnetic resonance elastograpy (MRE) has the potential to address this fundamental limitation by providing information on regional tissue mechanical properties.

Challenges to Visualizing Shear Waves within the Lung

As with all other organs, MRE requires as a first step the ability to clearly visualize propagating shear waves within the phase-difference MR images acquired during and MRE experiment. Unlike solid organs such as the liver, the lungs pose unique challenges that can be categorized into three distinct groups that include;

Tissue Density and Susceptibility Induced Signal Loss

Solid organs such as the liver and muscle are generally assumed to have a uniform physical density approximately equal to 1,000 kg/m^3. Because of the highly porous nature of the lung, the density of lung is considerably lower, ranging from approximately 400 to 900 kg/m^3 for normal and severely edematous lung respectively [4]. The direct consequence of this decrease in physical density is an appropriate scaling of the native MR signal. For a normal lung at least a 2.5 fold decrease in the MR signal can be expected due to physical density changes alone. Throughout the respiratory cycle, lung density can also be expected to change by a further factor of approximately two [5]. There are two direct consequences of this decrease in physical density, the first being an appropriate decrease in the MR signal, as measured by the signal-to-noise ratio (SNR), and secondly there is a need to apply an estimate of tissue density that reflects the actual density of the lung under interrogation. In general, higher MR SNR is sought to either improve spatial resolution or decrease imaging time. However, because MRE is a phase-contrast-based method, calculation of shear modulus is relatively insensitive to loss of SNR of the magnitude information as long as sufficient phase coherence can

be maintained so as to accurately encode the propagating shear wave. Because lung density cannot be assumed to be constant throughout the respiratory cycle or equal to that of solid organs (1,000 kg/m^3), a more accurate estimate must be obtained if shear stiffness estimates are to represent the intrinsic mechanical properties of the lung. A first order approximation of a global density less than 1,000 kg/m^3 can be obtained by referring to prior published values for normal and diseased lungs while regional estimates can be obtained from MR-based measurements using MR imaging techniques such as T_2 or T_2^* weighted images [6].

Although a decrease in the physical density of the lung will compromise the SNR of the MR signal, susceptibility induced signal loss, as measured by the T_2^* relaxation parameter, is the dominant factor that, in comparison to solid organ MR imaging makes lung MR signal starved. The complex organizational structure of the lung whose design is optimized to provide maximum surface area between air spaces and lung parenchyma necessary for gas exchange results in a T_2^* of the order of 1–2 ms [7], a 40-fold decrease compared to the typical T_2 value of lung of 80 ms [8]. Coupled with a relatively long T_1 relaxation value of ~830 ms [8], the lung is suited to neither spin-echo or gradient-echo based pulse sequences. Lung imaging with gradient-echo pulse sequences is restricted to those sequences in which the TE is of the order of the T_2^* of lung so as to maintain enough SNR to encode the resultant image echo (for a T_2^* of 2 ms, ~63 % of the signal will be lost for a TE of 2 ms). Spin-echo based techniques can be applied to overcome some of the limitations of an ultrashort T_2^* with the provision that the TE of the imaging sequence must be sufficiently short in order to overcome intravoxel phase dispersion and the subsequent formation of an image echo. The relatively long T_1 of the lung also requires a longer TR of the imaging sequence which has implications for overall scan time and hence the ability to acquire sufficient MRE phase offset data within a typical breath hold. Techniques to address the limitations imposed as a result of ultra short T_2^* values will be discussed in proceeding sections.

Volumetric Changes and Cardiac Motion

Volumetric changes of the lung throughout the respiratory cycle, coupled with cardiac motion represent two unique challenges to performing MRE. For normal adults, the volume of the lung can increase from the residual volume by at least 4.5 L until total lung capacity has been reached [1]. Not only does this result in a decrease in physical density (as well as a shortening of the T_2^*) of parenchyma but it also alters the intrinsic mechanical properties of the lung. Ex vivo animal studies have demonstrated that both the bulk and shear modulus are linearly related to transpulmonary pressure, P_{tp} and that both increase with increasing pressure [9–11]. The consequence of an increasing shear modulus with inflation pressure means that, in order to provide an accurate and reproducible estimate of shear modulus, the MRE acquisition sequence must be performed at a constant value of P_{tp} by using either some form of respiratory gated imaging method or by performing imaging within a single breath hold. In addition, the effects of respiration on shear wave propagation remains generally unknown. Cardiac motion introduces an additional source of signal perturbation due to the effects of both pulsatile blood flow as well as bulk motion of the heart itself. These two effects introduce ghosting artifacts and signal loss that affects both the magnitude and phase of the MR signal within the lung. The effect of cardiac motion is demonstrated in Fig. 9.1, which shows both axial magnitude and phase spin-echo MRE images of a porcine lung before and after sacrifice. While the imaging parameters are identical, Fig. 9.1a was acquired when cardiac motion was present while Fig. 9.1b was acquired without

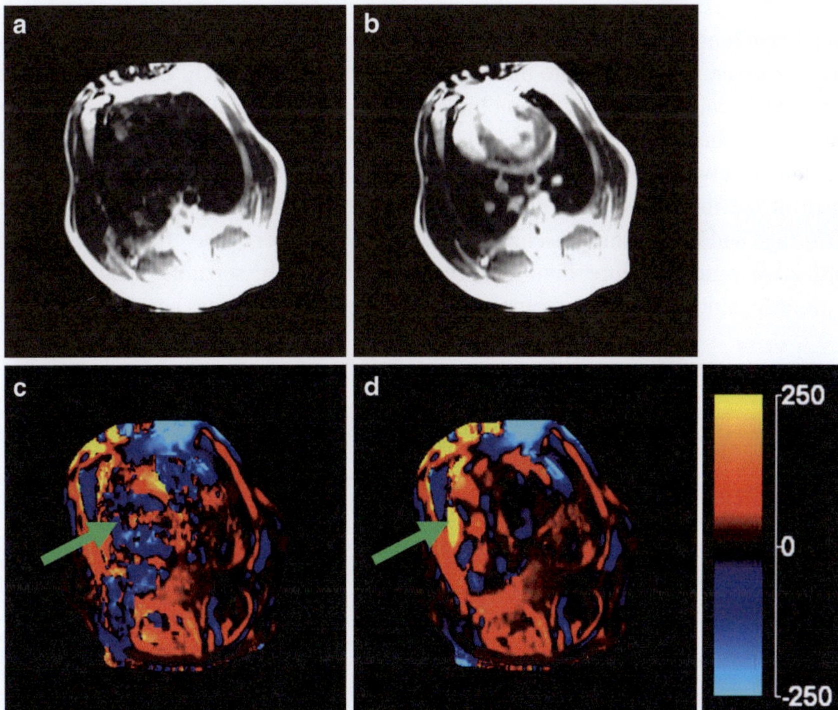

Fig. 9.1 In vivo axial spin-echo MRE magnitude and phase-difference data obtained from an anesthetized pig before (**a, c**) and immediately following (**b, d**) expiration. The *green arrow* shows the shear wave within the lung, emanating from the anterior chest wall and can be clearly identified when cardiac motion is absent (**d**). In the presence of cardiac motion, severe ghosting corrupts the phase-difference image (**c**)

cardiac motion. In both the magnitude and phase images of (a), ghosting and signal loss are present while in the absence of this effect, the heart is easily visualized and more importantly, the magnitude and phase images are artifact free. The presence of such severe artifacts suggests that some form of cardiac triggered MRE data acquisition will be necessary in order to acquire lung MRE data in vivo.

Shear Wave Driver Technology

The thoracic cavity, rib cage, and pleural space pose unique challenges to the generation of uniform plane shear waves within the lung. Successful translation of MRE from the research environment into routine clinical practice will most likely be achieved through the use of so-called passive drum drivers in which the vibration of a flexible membrane on the surface of the driver in contact with the skin of the patient produces the necessary shear wave by means of mode conversion of longitudinal waves. Ideally, as little as possible tissue should separate the organ under interrogation and the driver membrane. Additionally, the tissue should be able to act as an effective couple between the driver and organ. The presence of the thoracic cavity, defined by the boney rib cage and intercostal muscles represents an effective rigid structure that serve to absorb a significant amount of energy transmitted

from the driver face. The presence of the pleural space filled with pleural fluid further attenuates vibrations. A considerable effort is currently underway to investigate novel methods of shear wave generation as well as driver designs to overcome these limitations. Despite these challenges, Fig. 9.1b demonstrates that, even with existing passive driver technologies, shear waves can be generated within the lung in vivo.

Conventional (^{1}H) MRE Methods

The application of ^{1}H imaging for lung MRE has several advantages when compared to those techniques that employ exogenous contrast agents (see proceeding section) including the availability of numerous MRE pulse sequences that require relatively minor modifications and the lack of additional specialized hardware beyond the MRE driver technology used for shear wave generation.

As described previously, the ultra short T_2^* of lung parenchyma imposes limitations on the type of imaging sequence as well as specific parameters chosen for lung MRE. Consequently, the majority of work published thus far has focused on the development of spin-echo based ^{1}H MRE imaging techniques. Figure 9.2 shows the first in vivo ^{1}H spin-echo MRE lung images acquired from a normal volunteer. The four phase offsets were acquired during separate breath holds of 32 s while each phase offset was acquired with the following

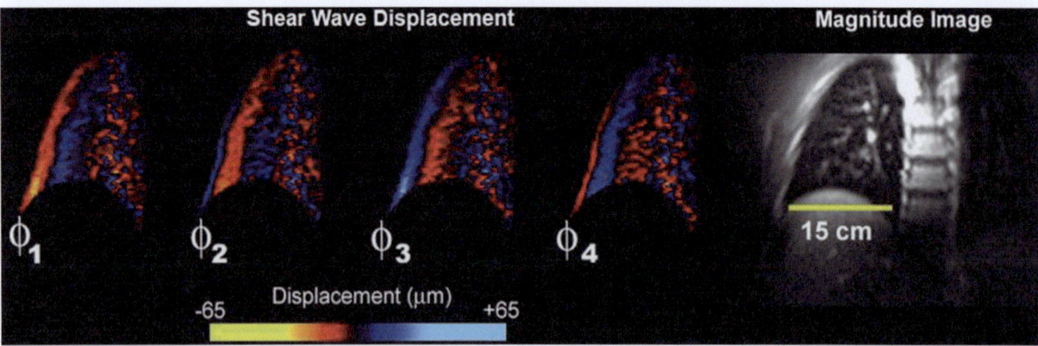

Fig. 9.2 Magnitude and phase-difference images acquired from a spin-echo MRE pulse sequence in a normal volunteer. The phase-difference images obtained at four equally spaced temporal offsets show the propagation of the shear wave from the lateral chest wall towards midline. Data reproduced from Goss et al. [12]

imaging parameters: TR/TE=250/17 ms, k_x/k_y= 128/128, and a receiver bandwidth of 64 kHz. Shear waves were generated from a passive driver located at the lateral chest wall at a frequency of 40 Hz and can be seen propagating away from the chest wall towards midline [12].

The choice of imaging parameters (i.e., TR and TE) for SE MRE applications is dictated by several tradeoffs. The choice of TR is governed by the balance between sufficient time for T_1 signal recovery (choice of longer versus shorter TR) and sufficiently short imaging time to allow breath hold acquisitions (shorter versus longer TR). Similarly, the choice of TE is determined by the compromise between the need for a short as possible TE value to minimize intravoxel phase dispersion and signal loss and a longer TE in order to encode cyclic displacements at frequencies between the tens to hundreds of Hertz. In general, the choice of TR is based on the need to reduce overall imaging time. While the magnitude of the MR signal is affected by choice of TR, the phase of the signal, in which shear wave displacements are encoded is less so and as such shorter TRs are generally preferred. For both ex vivo, in vivo animal and human studies [12, 13], TRs of between 200 and 300 ms are typically used, making these imaging sequences highly T_1-weighted. The overriding consideration for the choice of TE value is to make this parameter as short as possible. TEs can be minimized by choosing a driving frequency of the passive driver system to be several hundred Hertz as opposed to the typical value for the majority of abdominal MRE applications which is of the order of 40-60 Hz. With a driving frequency of 200 Hz, a TE of 22 ms can be achieved [13]. However, a disadvantage of the use of these relatively high frequencies is the decreased efficiency of generating longitudinal sound waves necessary to vibrate the passive driver as well as the increased attenuation of mode converted shear waves at these higher frequencies resulting from the viscous properties of tissue. The former problem can be partially overcome by increasing the driving voltage of the audio speaker used to generate the longitudinal sound waves. However, this poses increased performance requirements both

in terms of the audio speaker as well as driving amplifier. In conventional spin-echo MRE pulse sequences, the interval between the 90° and 180° RF pulse is typically used to encode a single period of the shear wave motion. Thus, since shear frequency is inversely proportional to period, higher frequencies will reduce the period or time interval necessary to encode shear wave induced displacements and hence produce a lower TE.

To address the limitations of existing spin-echo MRE pulse sequences new, novel methods are currently being developed to both reduce the frequency of the longitudinal wave source while simultaneously reducing TE. These methods include reduction of the motion encoding gradient (MEG) duration and the incorporation of the MEGs into the crusher gradients which are typically played out on either side of the slice selection gradient of the 180° refocusing pulse. By incorporating the MEGs in this manner, the TE of the imaging sequence can be decreased significantly compared to the conventional spin-echo MRE pulse sequence. This is demonstrated in Fig. 9.3a, b which shows a conventional spin-echo based MRE pulse sequence with a bipolar 16.66 ms MEG pair with a TE of 44 ms and a modified spin-echo MRE pulse sequence with a TE of 10 ms respectively. The pulse sequence shown in 3a has the maximum sensitivity for 60 Hz shear motion. To achieve the pulse sequence design shown in 3b, the bipolar MEG pair was split into two 8.33 ms MEG lobes, each placed on either side of the 180° pulse. In this example the MEG lobes act as both motion encoding and crusher gradients.

Hyperpolarized Gas Imaging of Lung Parenchyma

A novel method of addressing the limitations imposed by the ultra short T_2^* of lung parenchyma is to introduce an exogenous contrast agent in which the T_2^* of the agent is much longer than that of the surrounding parenchyma. Under these circumstances either gradient or spin-echo imaging sequences can be used for lung MRE. One such family of agents are Noble

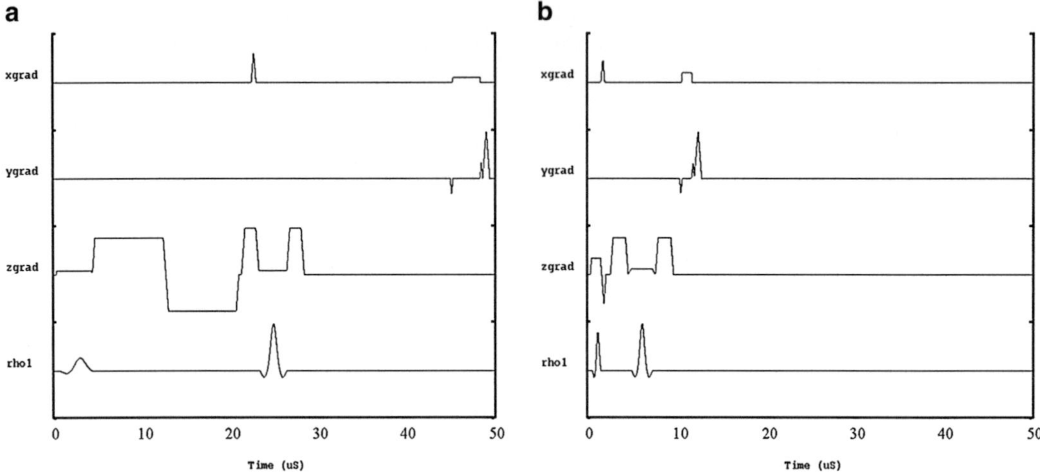

Fig. 9.3 (**a**) Conventional spin-echo based MRE pulse sequence with 16.66 ms bipolar MEG pair and a killer gradient on each side of the refocusing pulse, having a TE of 48 ms. (**b**) Modified MRE pulse sequence with two, 2 ms gradient lobes acting as MEG as well as a killer gradient having a TE of 10 ms

gas isotopes such as ^{3}He and ^{129}Xe which have been successfully employed to evaluate a variety of lung MR imaging applications including static and dynamic perfusion, diffusion [14], ventilation [15], and response to methacholine and exercise challenge [16]. Because these agents have an odd atomic mass number, they have a net magnetic moment that can be used as a contrast agent, similar to conventional ^{1}H imaging. However, such agents typically have a relatively low net magnetic polarization under thermal (i.e., room temperature) conditions. The small signal resulting from the gas can be overcome by the process of hyperpolarization in which the polarization of the gas is increased from approximately 1×10^{-6} to approximately 10^{-1} [17]. Hyperpolarization is complex and expensive and as such, the concept of a centralized polarization facility has been developed to distribute the gas to sites that do not possess the necessary polarization hardware [17].

While hyperpolarized Noble gases can be used for both gradient and spin-echo imaging, the majority of applications have involved ultra fast gradient-echo imaging sequences using very small flip angles (<10°). This is because the magnetization achieved by means of hyperpolarization is non-renewable. The use of low flip angle gradient-echo pulse sequences ensures that sufficient magnetization will be available throughout the acquisition of all image echoes necessary for image formation. The use of ultra fast gradient-echo sequences has an additional advantage in that imaging can be performed on the order of several seconds, well within the breath hold time for some of the most ambulatory patients.

Figure 9.4 shows an example of performing MRE using a gradient-echo based MRE pulse sequence using hyperpolarized ^{3}He in an ex vivo porcine lung. In this instance the imaging parameters were as follows; TR/TE = 100/8.3 ms, $k_x/k_y = 128/64$, bandwidth = 32 kHz, and field of view of 20 cm [13]. The presence of a shear wave emanating from the driver plate, as seen in the phase-difference images demonstrates that diffusion of the hyperpolarized gas within the respiratory zone of the lung is sufficiently restrictive so as to allow coherent phase accumulation of the magnetization of the gas and thus encode the displacement of the shear wave as it propagates throughout the lung. This suggests that hyperpolarized Noble gases have the potential for use as a contrast agent for lung MRE and that, while gases do not support shear, shear wave displacements can be encoded into the magnetization of a hyperpolarized gas because of gas trapping within the respiratory zone of the lung. The reconstructed shear wave map or elastogram also demonstrates regional changes in shear

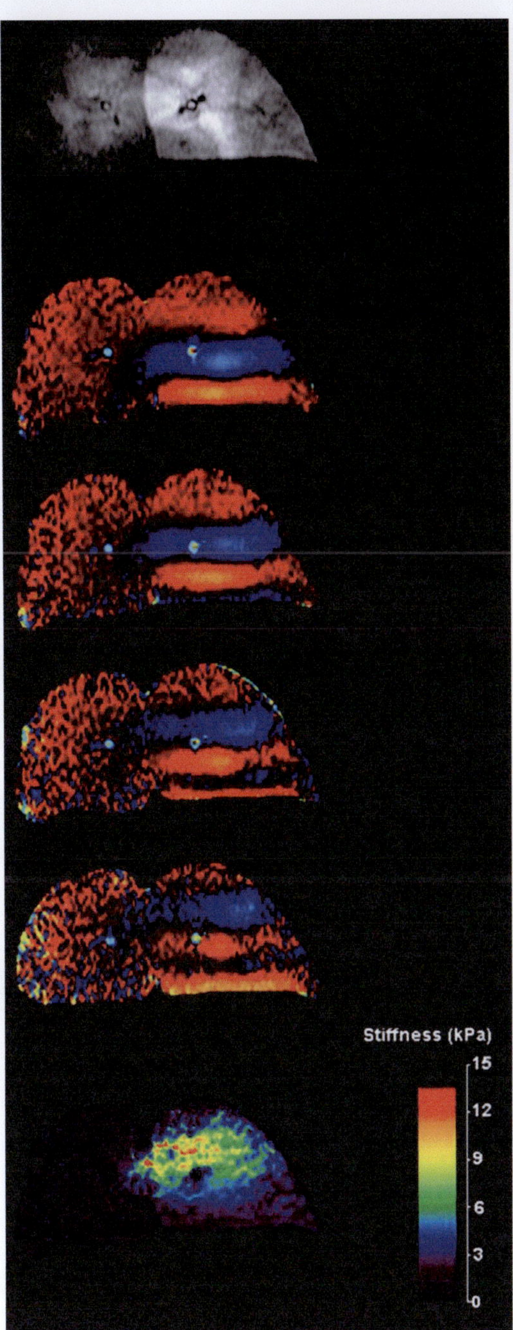

Fig. 9.4 Magnitude, phase difference, and reconstructed elastogram of an ex vivo porcine lung infused with hyperpolarized ^{3}He and imaged with a gradient-echo MRE pulse sequence. The phase-difference images are acquired at four different temporal offsets and show the propagation of the shear wave from the driver plate in the lung that is sitting on the driver plate. In contrast, the opposite lung in which no mechanical energy is supplied do not show signs of shear waves. The reconstructed elastogram shows increased stiffness about the bronchi, indicating that gas is perfusing into the lung. Data reproduced from McGee et al. [13]

modulus throughout the lung, most likely due to the relatively higher pressure of the regions of the lung about the bronchi present within the imaging slice, from which the gas is perfusing from.

An important consideration in performing MRE of the lung using a hyperpolarized gas is the effect diffusion induced signal loss as a result of application of the MRE MEGs. Diffusion induced signal loss is described by the equation: $S/S_o = e^{-b \cdot \mathrm{ADC}}$ where S/S_o is the ratio of the signals with and without diffusion sensitized gradients, b the so-called b-factor which is a function of the diffusion gradient waveform and the ADC or apparent diffusion coefficient [8]. MRE MEGs are similar to diffusion gradient waveforms because they are designed to accumulate zero phase for static tissue and as such, depending upon the amplitude of the MEGs, MRE images can be considered highly diffusion weighted. This is especially true for highly mobile small molecules such as ^{3}He in the gaseous form. To appreciate the significance of this affect, consider the difference in diffusion between unbound water at 25 °C (ADC $= 2.4 \times 10^{-5}$ cm^2/s [8]) versus that of ^{3}He. In vivo calculations of the ADC in normal volunteers of ^{3}He have estimated this value to be approximately 0.25 cm^2/s [18, 19] or $\sim 10^5$ times greater than extracellular water. For a single bipolar rectangular MEG MRE waveform added to a GRE pulse sequence, the b value is given by the relationship: $b = 2\gamma^2 G^2 \delta^3 / 3$ where γ is the gyromagnetic ratio, G the amplitude of the gradient waveform, and δ the temporal duration of a single gradient lobe [8]. Figure 9.5 is a two-dimensional plot of the ratio S/S_o for a rectangular MRE motion encoding waveform as a function of the amplitude, G, and frequency of the waveform. The rectangle indicates the zone in which MRE experiments are typically performed. That is, gradient amplitudes between 5 and 17.5 mT/m and motion sensitization frequencies between 75 and 250 Hz. Within this zone the signal loss due to diffusion ranges from 10 to 90 %. While this signal loss will reduce the SNR of the magnitude data, it is the phase image that contains the shear displacement information. Therefore, careful choice of MRE parameters will ensure that the measurement of cyclic coherent motion due to wave propagation is independent of the effects of diffusion.

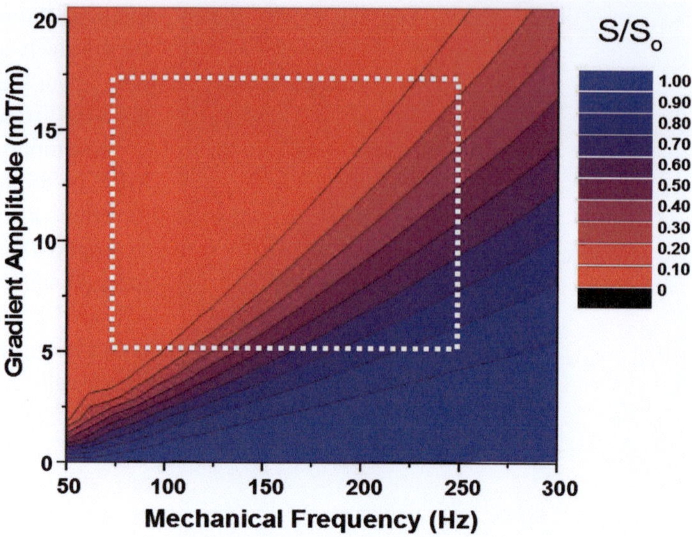

Fig. 9.5 MRE induced signal loss ratio (S/S_o) as a function of gradient amplitude and mechanical frequency of the waveform. The rectangular region includes the "normal operating zone" for MRE of the lung

Future Directions

Despite the technical challenges, lung MRE continues to be an active field of research. Two such areas involve the development of more accurate models of lung density and improved inversion algorithms that more closely describe shear wave propagation within the honeycomb-like structure of lung parenchyma.

Estimating Regional Density Variations

Because lung parenchyma is an inherently porous elastic medium comprised of air-filled voids on the micro scale, the assumption of a constant density within the medium is no longer valid even at this resolution. For example, if we are able to obtain elastic displacements with an isotropic spatial discretization of one millimeter, it has been shown that one cubic millimeter of lung parenchyma contains around 170 alveoli [20]. Despite its porous nature, an "effective" density can be derived by the superposition of the local microscopic densities. This can be expressed more formally as $\rho(r) = \dfrac{1}{N} \sum_{i=1}^{N} \rho_{0i}(r - r_{0i})$ where we assume that each $\rho_{0i}(r - r_{0i})$ is a normalized local density. That is, if a measure of regional, macroscopic density can be obtained, it is related to the microscopic regional density values by linear summation.

This regional density measure can be utilized in the equations of motion, providing a regional measure of shear stiffness that includes regional density variations. By rearranging the equation of motion for shear wave propagation, the shear modulus, μ_m can be rearranged so that

$$\mu_m = \frac{-\int \omega^2 \rho u_3^T v_3 \, d\Omega_m}{\int u_3^T \left[\dfrac{\partial^2 v_3}{\partial x_1^2} + \dfrac{\partial^2 v_2}{\partial x_2^2} \right] d\Omega_m} \quad \text{where } \rho,\ u_3^T \text{ and } v_3$$

are now functions of position. In this case, the quantity μ_m now represents a solution for an "effective" macroscopic shear modulus representative of the three-dimensional volume of the local window, Ω_m, based on the displacements, their derivatives, and local density variations. Advanced lung MRE inversion algorithms will require inclusion of regional density values using this or other models that account for the porous, regionally heterogeneous characteristics of lung parenchyma physical density.

Equations of Motion in Heterogeneous Porous Materials

In addition to regional estimates of density, the next generation of lung MRE inversion algorithms will need to accurately model the porous nature of the lung.

$$\sigma = \left(P - 2N\right)\nabla \cdot uI + N\left[\nabla u + \left(\nabla u\right)^{T}\right] + Q\nabla \cdot UI \quad (1)$$

$$s = Q\nabla \cdot u + R\nabla \cdot U \quad (2)$$

$$\rho_{11}\frac{\partial^2 \vec{u}}{\partial t^2} + \rho_{12}\frac{\partial^2 \vec{U}}{\partial t^2} = \nabla \cdot \sigma + b\frac{\partial}{\partial t}\left(\vec{U} - \dot{u}\right) \quad (3)$$

$$\rho_{12}\frac{\partial^2 \vec{u}}{\partial t^2} + \rho_{22}\frac{\partial^2 \vec{U}}{\partial t^2} = \nabla s - b\frac{\partial}{\partial t}\left(\vec{U} - \vec{u}\right) \quad (4)$$

As has been demonstrated by Biot [21], a porous medium is one which can support three wave types: a fast compressional wave, a slow compressional wave, and a slow shear wave. These wave types are a consequence of the interaction of the coupled vibrational behavior of the solid skeletal background matrix and the encased fluid in the pores. Using the notation of Biot [21] it is possible to define the stress-strain relations as described in Eqs. (1) and (2) with equations of motion given by Eqs. (3) and (4). The vectors $\vec{u}\left(x,t\right)$ and $\vec{U}\left(x,t\right)$ are equal to the solid and fluid displacements, respectively; σ is the solid stress tensor; s is a scalar proportional to the fluid pressure; and I is the second-order identity tensor. The densities ρ_{ij} in Eqs. (3) and (4) represent the inertia of the two phases and are related to the fluid and solid densities ρ_f and ρ_s which are given as $\rho_{11} + \rho_{12} = (1 - \phi)\rho_s$ and $\rho_{12} + \rho_{22} = \phi\rho_f$ where ϕ is the porosity or volume fraction of the fluid phase. The coupling mass $\rho_{12} < 0$ is sometimes written as $\rho_{12} = -(T - 1)\phi\rho_f$, where the tortuosity

$T > 1$ depends upon the pore geometry. The parameter b in Eqs. (3) and (4) represents the resistive damping due to relative motion between the fluid and solid. It incorporates viscous dissipation into the equations of motion and, therefore, is a major source of attenuation in waves.

While implementation of the model proposed by Biot [21] is complex and challenging, these efforts are obviously necessary in order to more accurately model the intrinsic mechanical properties of the lung. These improvements will further our understanding of both normal lung function as well as those disease processes that affect lung parenchyma such as ILD. These challenges are significant, yet the potential to impact patient care and disease management is much greater.

Pulse Sequence Developments

Building upon previous technical developments using spin-echo pulse sequences for lung MRE, Mariappan et al. [22] have demonstrated that a spin-echo, echo planar imaging (EPI) sequence can be used to spatially resolve the shear stiffness of lung parenchyma in both normal volunteers and a single patient with known ILD. In this work, the authors obtained a two-fold decrease in image acquisition time compared to their prior spin-echo MRE pulse sequence. The spin-echo EPI sequence was able to acquire four phase offset images of the shear wave field encoding a single direction of motion within the lung in a total acquisition time of 30 s, which was split into two 15 s breath holds. The authors also included a 2D density mapping method thereby allowing absolute estimates of parenchymal shear stiffness to be obtained. It is expected that additional technical advances such as encoding all three physical directions of motion so as to fully sample the complex shear wave field will be required to further advance this technique into routine clinical practice. However, this work demonstrates that significant improvements, in this case a reduction in acquisition time, can be achieved using previously developed MR pulse sequence techniques.

References

1. Levitzky MG. Pulmonary physiology. 7th ed. New York: McGraw-Hill Medical; 2007.
2. Suki B, Ito S, Stamenovic D, Lutchen KR, Ingenito EP. Biomechanics of the lung parenchyma: critical roles of collagen and mechanical forces. J Appl Physiol. 2005;98(5):1892–9.
3. Suki B, Bates JH. Extracellular matrix mechanics in lung parenchymal diseases. Respir Physiol Neurobiol. 2008;163(1–3):33–43.
4. Bai R, Yu X, Wang D, Lv J, Xu G, Lai X. The densities of visceral organs and the extent of pathologic changes. Am J Forensic Med Pathol. 2009;30(2): 148–51.
5. Kohda E, Shigematsu N. Measurement of lung density by computed tomography: implication for radiotherapy. Keio J Med. 1989;38(4):454–63.
6. Hopkins SR, Henderson AC, Levin DL, Yamada K, Arai T, Buxton RB, Prisk GK. Vertical gradients in regional lung density and perfusion in the supine human lung: the Slinky effect. J Appl Physiol. 2007;103(1):240–8.
7. Hatabu H, Alsop DC, Listerud J, Bonnet M, Gefter WB. T2* and proton density measurement of normal human lung parenchyma using submillisecond echo time gradient echo magnetic resonance imaging. Eur J Radiol. 1999;29(3):245–52.
8. Bernstein MA, King KF, Zho XJ. Handbook of MRI pulse sequences. New York: Elsevier Academic Press; 2004. p. 1017.
9. Lai-Fook SJ, Hyatt RE, Rodarte JR. Elastic constants of trapped lung parenchyma. J Appl Physiol. 1978;44(6):853–8.
10. Lai-Fook SJ, Hyatt RE. Effects of age on elastic moduli of human lungs. J Appl Physiol. 2000;89(1):163–8.
11. Stamenovic D, Yager D. Elastic properties of air- and liquid-filled lung parenchyma. J Appl Physiol. 1988; 65(6):2565–70.
12. Goss BC, McGee KP, Ehman EC, Manduca A, Ehman RL. Magnetic resonance elastography of the lung: technical feasibility. Magn Reson Med. 2006;56(5):1060–6.
13. McGee KP, Hubmayr RD, Ehman RL. MR elastography of the lung with hyperpolarized 3He. Magn Reson Med. 2008;59(1):14–8.
14. Moller HE, Chen XJ, Saam B, Hagspiel KD, Johnson GA, Altes TA, de Lange EE, Kauczor HU. MRI of the lungs using hyperpolarized noble gases. Magn Reson Med. 2002;47(6):1029–51.
15. Altes TA, Powers PL, Knight-Scott J, Rakes G, Platts-Mills TA, de Lange EE, Alford BA, Mugler 3rd JP, Brookeman JR. Hyperpolarized 3He MR lung ventilation imaging in asthmatics: preliminary findings. J Magn Reson Imaging. 2001;13(3):378–84.
16. Samee S, Altes T, Powers P, de Lange EE, Knight-Scott J, Rakes G, Mugler 3rd JP, Ciambotti JM, Alford BA, Brookeman JR, Platts-Mills TA. Imaging the lungs in asthmatic patients by using hyperpolarized helium-3 magnetic resonance: assessment of response to methacholine and exercise challenge. J Allergy Clin Immunol. 2003;111(6):1205–11.
17. van Beek EJ, Schmiedeskamp J, Wild JM, Paley MN, Filbir F, Fichele S, Knitz F, Mills GH, Woodhouse N, Swift A, Heil W, Wolf M, Otten E. Hyperpolarized 3-helium MR imaging of the lungs: testing the concept of a central production facility. Eur Radiol. 2003;13(12):2583–6.
18. Fain SB, Altes TA, Panth SR, Evans MD, Waters B, Mugler 3rd JP, Korosec FR, Grist TM, Silverman M, Salerno M, Owers-Bradley J. Detection of age-dependent changes in healthy adult lungs with diffusion-weighted 3He MRI. Acad Radiol. 2005;12(11):1385–93.
19. Morbach AE, Gast KK, Schmiedeskamp J, Dahmen A, Herweling A, Heussel CP, Kauczor HU, Schreiber WG. Diffusion-weighted MRI of the lung with hyperpolarized helium-3: a study of reproducibility. J Magn Reson Imaging. 2005;21(6):765–74.
20. Ochs M, Nyengaard JR, Jung A, Knudsen L, Voigt M, Wahlers T, Richter J, Gundersen HJ. The number of alveoli in the human lung. Am J Respir Crit Care Med. 2004;169(1):120–4.
21. Biot MA. Theory of propagation of elastic waves in a fluid-saturated porous solid. I: higher frequency range. J Acoust Am. 1956;28(2):179–91.
22. Mariappan YK, Glaser KJ, Levin DL, Vassallo R, Hubmayr RD, Mottram C, Ehman RL, McGee KP, Estimation of the absolute shear stiffness of human lung parenchyma using 1H spin echo, echo planar MR elastography. J. Magn. Reson. Imaging. 2013; doi: 10.1002/jmri.24479

Magnetic Resonance Elastography of the Heart

Kiaran P. McGee, Arunark Kolipaka, Philip Araoz, Armando Manduca, Anthony Romano, and Richard L. Ehman

Anatomy, Physiology, and Mechanical Basis of Cardiac Function, Disease Initiation, and Progression

The heart is a four chambered, hollow organ that is the size of a human fist and weighs approximately one pound. Surrounded by the two lungs, it sits obliquely within the thoracic cavity with its apex resting on and tethered to the diaphragm slightly to the left of midline while its broader base from which the great vessels emerge is located approximately below the sternum. The walls of the four chambers of the heart are comprised primarily of twisted cardiac muscle bundles that form ring-like arrangements and aid in the mechanical contraction of the chambers.

The function of the heart is to provide the mechanical force to circulate blood throughout the vascular system, providing a high pressure source for oxygenated blood (circulatory system) and low pressure for deoxygenated blood (venous system) return. Cardiac function, as measured by the volume of blood pumped by the heart within a single heart beat (stroke volume) is intrinsically related to cardiac muscle tone and contractility. This parameter is commonly quantified by measurement of myocardial elasticity or stiffness. The term myocardial stiffness is somewhat generic and has evolved alongside those technologies that have attempted to quantify this parameter. Within the context of cardiac MRE, the term will be used to describe the shear modulus.

Existing Methodology for Studying Myocardial Mechanical Properties

The current gold standard for assessing myocardial stiffness is by means of invasive pressure–volume (PV) loops obtained throughout the cardiac cycle. The process involves insertion of a pressure catheter into the femoral artery which is then snaked into the left ventricle (LV) of the heart.

K.P. McGee, Ph.D. (✉) • A. Romano, Ph.D.
United States Naval Research Laboratory,
Washington, DC, USA
e-mail: McGee.Kiaran@mayo.edu;
romano.anthony@mayo.edu

A. Kolipaka, Ph.D.
Department of Radiology, 395 W 12th Ave.,
Columbus, OH, USA

Department of Radiology, The Ohio
State University, Colombus, OH, USA
e-mail: kolipaka.arum@mayo.edu

P. Araoz, M.D. • A. Manduca, Ph.D.
R.L. Ehman, M.D.
Department of Radiology, Mayo Clinic
College of Medicine, 200 First Street SW,
Rochester, MN 55905, USA
e-mail: Araoz.philip@mayo.edu; manduca.
armando@mayo.edu; ehman.richard@mayo.edu

S.K. Venkatesh and R.L. Ehman (eds.), *Magnetic Resonance Elastography*,
DOI 10.1007/978-1-4939-1575-0_10, © Springer Science+Business Media New York 2014

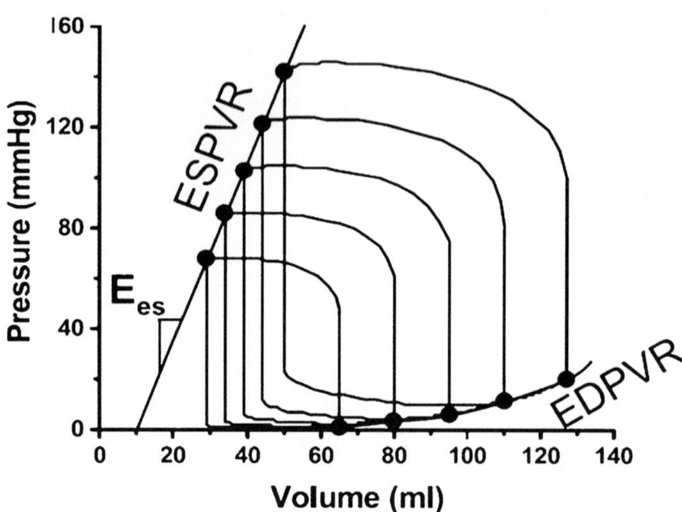

Fig. 10.1 Measurement of the EDPVR requires creating a family of pressure–volume (PV) loops, which are obtained by simultaneously measuring pressure and volume at different loading conditions. End-diastolic pressure points are plotted to create the EDPVR curve, the slope of which is an estimate of chamber stiffness. In these hypothetical PV loops, the end-systolic points have also been plotted to obtain the end-systolic pressure volume relationship (ESPVR), the slope of which is E_{es}. Reprinted from Burkhoff et al. [1]

The cavity pressure throughout the cardiac cycle is measured and plotted against the volume of the ventricle, most commonly determined from ultrasound-based measurements. These PV measurements must be made at multiple loading conditions (often with inferior vena cava occlusion) to create a family of PV loops. From these loops, the end-diastolic pressure–volume points are plotted in order to define the end-diastolic pressure volume relationship (EDPVR) [1] as demonstrated in Fig. 10.1, the slope of which (dP/dV) is referred to as the LV chamber stiffness. This term is incorporated into mathematical models designed to assess chamber stiffness. One such model proposed by Mirsky and Parmley [2] is shown in Eqs. (10.1) and (10.2) and assumes; (1) spherical geometry of the LV cavity and myocardium, (2) use of instantaneous stresses and strains at the mid wall for the evaluation of elastic stiffness, and (3) the LV wall is incompressible (i.e., the wall volume remains constant throughout the cardiac cycle) [2]. This model calculates the Young's modulus E according to the relationship:

$$E = 3\sigma_m \left[\frac{\left(1+\left(\dfrac{V_w}{V}\right)\left(\dfrac{a^2}{a^2+b^2}\right)\right)\left(1+\left(\dfrac{V}{P}\right)\left(\dfrac{dP}{dV}\right)\right) - }{\left(\left(\dfrac{b^2}{a^2+b^2}\right)\left(\dfrac{b^3-a^3}{2R^3+b^3}\right)\right)} \right] \quad (10.1)$$

$$\sigma_m = P\left(\frac{V}{V_w}\right)\left(1+\frac{b^3}{2R^3}\right) \quad (10.2)$$

where σ_m = stress at the mid wall, V = volume of the chamber (i.e., LV cavity), V_w = volume of the wall (remains constant as the material is assumed to be incompressible), P = internal cavity pressure (i.e., chamber pressure), a = inner radius of the shell, b = outer radius of the shell, $R = (a+b)/2$ (i.e., mid wall radius), and dP/dV = chamber stiffness [1, 2].

Because the EDPVR is non-linear, there are a variety of curves to which it may be fit and there is currently no consensus regarding which is the most accurate.

MRE Pulse Sequences for Cardiac MRE Applications

Accurate assessment of myocardial stiffness using MRE-based methods is challenging. Respiratory induced bulk motion as well as the volumetric changes of the heart's chambers as they expand and contract throughout the cardiac cycle violate the assumptions imposed on existing MRE inversion algorithms and thus can produce erroneous estimates of myocardial stiffness. The unique geometry of the heart also means that wave propagation cannot be described as a plane shear wave propagating within an infinite medium. Without appropriate modeling of the boundary conditions imposed by the geometry of the heart and the resultant effect on wavelength, erroneous assessment of myocardial stiffness can result. The following discussion describes the various methods under development to address these challenges.

Advanced Geometric Modeling of the Left Ventricle

Due to the unique geometry of the heart—four thin walled fluid filled cavities that undergo cyclic changes in wall thickness and volume—several underlying assumptions imposed by commonly used inversion algorithms become invalid and therefore create the potential for incorrectly estimating myocardial stiffness. To improve the accuracy of this estimate, it is necessary to develop inversion algorithms that more closely model the heart's physical characteristics. When vibrations from an external source are introduced into an object in which the wavelength of the vibration is greater than the thickness of the object, flexural (i.e., bending) waves are generated. Kolipaka et al. [3] have investigated flexural waves in bounded media such as thin beams, plates, and spherical shells as potential models for visualizing vibrations below several hundred Hertz within the myocardium from a passive longitudinal driver such as those used in clinical MRE studies to more accurately assess myocardial stiffness. Because spherical shells most closely match the geometry of the chambers of the heart, only this model will be presented.

When vibrations are introduced into a spherical shell, flexural waves are generated as a result of the shell's boundary conditions. By application of Hamilton's variation principle with the assumption of mid surface deflections and non-torsional axisymmetric motion, the equations of motion of these waves can be solved to provide an estimate of the flexural motion and are given in the equation below [3] assuming that; (1) the displacement of

the shell is small in comparison to its thickness, (2) the thickness of the shell is small compared with the smallest radius of curvature, (3) "fibers" or elements of the shell initially perpendicular to the middle surface remain so after deformation and are themselves not subject to elongation, and (4) the normal stress acting on planes parallel to the shell middle surface is negligible in comparison with other stresses.

$$\beta^2 \frac{\partial^3 u}{\partial \theta^3} + 2\beta^2 \cot\theta \frac{\partial^2 u}{\partial \theta^2} - \left[\begin{array}{c} (1+v)(1+\beta^2) \\ + \beta^2 \cot^2\theta \end{array} \right] \frac{\partial u}{\partial \theta}$$

$$+ \cot\theta \left[(2 - v + \cot^2\theta)\beta^2 - (1+v) \right] u - \beta^2 \frac{\partial^4 w}{\partial \theta^4}$$

$$- 2\beta^2 \cot\theta \frac{\partial^3 w}{\partial \theta^3} + \beta^2 (1 + v + \cot^2\theta) \frac{\partial^2 w}{\partial \theta^2}$$

$$- \beta^2 \cot\theta (2 - v + \cot^2\theta) \frac{\partial w}{\partial \theta} - 2(1+v)w$$

$$- \frac{a^2 \ddot{w}}{c_p^2} = -p_a \frac{(1-v^2)a^2}{Eh} \qquad (10.3)$$

This equation provides an estimate of Young's modulus, E from which the shear modulus can be obtained by substation in the equation $\mu = E/2(1+v)$ where v is the Poisson's ratio of the material.

Figure 10.2 shows both simulation and experimental displacement maps for the radial and circumferential components of the flexural wave resulting from the application of a sinusoidal force at a frequency of 200 Hz to the top surface of a thin silicone rubber spherical shell of 10 cm diameter and thickness of 1 cm. The in-plane components of displacement were encoded using a gradient-echo MRE sequence with 5 ms duration MEG. The third column of this figure shows the reconstructed map of shear modulus (i.e., the elastogram). There are several reasons for the relative heterogeneity of the elastogram images including the sensitivity of the above equation of motion to noise as a result of the third order derivatives as well as the singularities at the two poles of the shell ($\theta = 0°$ and $90°$). Despite these sources of error, the model is able to provide an accurate estimate of shear stiffness which has been validated using finite element modeling simulations and is shown in Fig. 10.3.

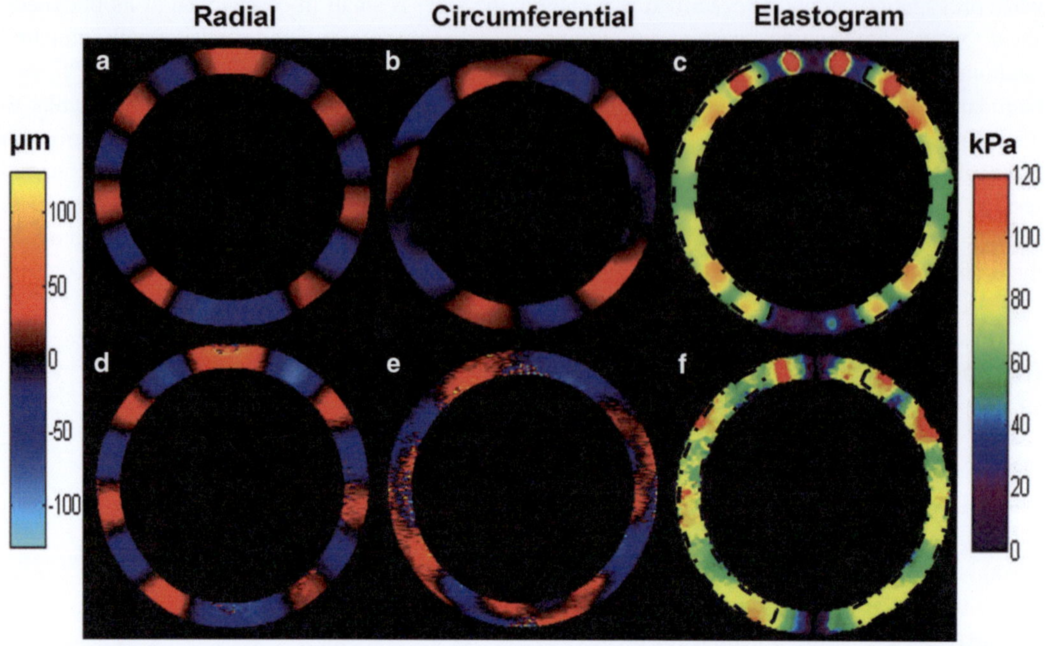

Fig. 10.2 Shear wave displacement images of a thin spherical shell undergoing mechanical excitation for both finite element modeling simulations (**a**, **b**) and MRE experiments (**d**, **e**), respectively, for a 1 cm thick shell vibrated at a frequency of 200 Hz. (**c**, **f**) The corresponding stiffness maps obtained from spherical shell inversion with a mean stiffness of 71.6 ± 14.3 and 71 ± 14 kPa respectively

Pressure–Volume Based Validation of Advanced Inversion Algorithms

Before MRE measures of myocardial stiffness can replace PV based estimates, the relationship between these two parameters must be determined. This relationship has been demonstrated in the work by Kolipaka et al. [4] who compared the shear stiffness derived from the PV model proposed by Mirsky [2] to MRE-derived estimates using the spherical shell equations above. Figures 10.4 and 10.5 are plots of shear stiffness for the two methods for a thin spherical shell phantom as a function of chamber pressure. Figure 10.4 was obtained when the phantom was inflated to a given static pressure while Fig. 10.5 is a plot of the two measures obtained when the cavity pressure varied under cyclic conditions. In both instances, correlation between the two methods was obtained indicating that MRE-based estimates of shear stiffness compare well to the current accepted gold standard of assessing myocardial function. More significantly, this

work also demonstrated that under conditions of both static and dynamic pressure, MRE-derived shear stiffness values were highly correlated ($R^2 = 0.9948$).

Cardiac MRE Pulse Sequences

Successful in vivo application of cardiac MRE must address the problems of respiratory and cardiac motion. Respiratory motion can be minimized by acquiring MRE data within a breath hold. However, the breath hold duration must be short enough so as to be tolerated by the most ambulatory of patients which, based on our own clinical experience should ideally be less than 20 s. Short breath holds are typically achieved in cardiac MR imaging by application of gradient-echo based pulse sequences such as spoiled or balanced gradient echoes. Preliminary cardiac MRE pulse sequence development within our own laboratory has focused on the adaptation of MRE techniques into cardiac spoiled gradient-echo

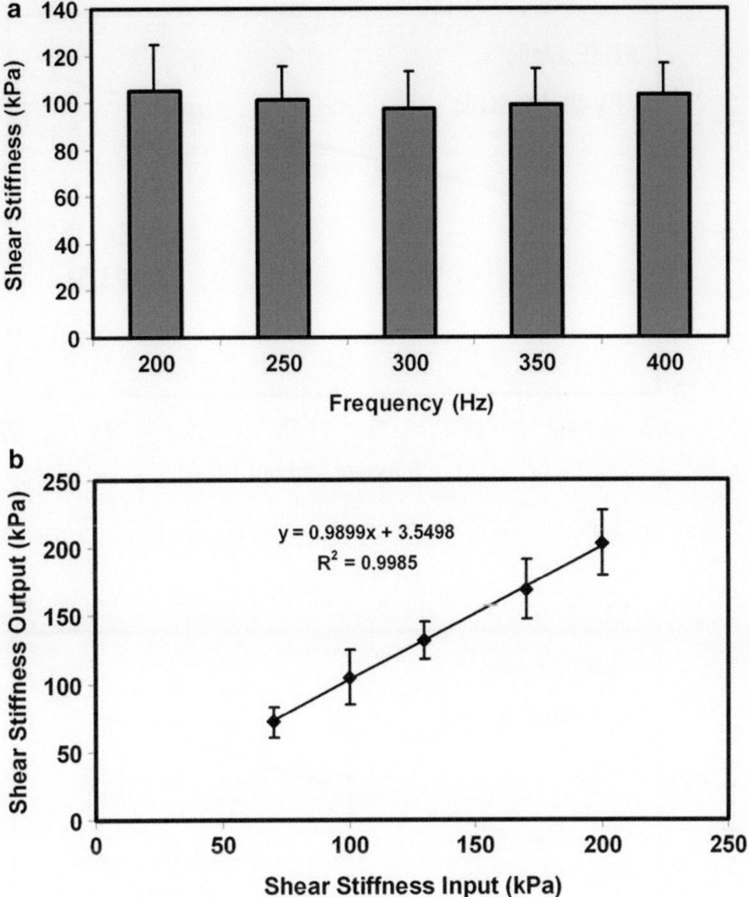

Fig. 10.3 (a) Shear stiffness plotted as a function of excitation frequency obtained from finite element modeling of a 1 cm thick shell experiencing excitation from a point source located at the north pole of the sphere for an input shear modulus of 100 kPa. (b) Shear stiffness of the spherical shell as an input into the finite element modeling of the thin spherical shell versus calculated shear modulus based on the finite element modeling displacement fields when input into the spherical shell wave field inversion algorithm. The shell was modeled with a thickness of 1 cm and an excitation frequency of 200 Hz

pulse sequences. A direct consequence of this approach is that existing, well developed cardiac MR sequences can be adapted to include MRE techniques. In addition, these sequences address the problem of cardiac motion by incorporation of ECG-gated segmented acquisition methods to effectively "freeze" cardiac motion and provide either cine or static images of the heart throughout the cardiac cycle.

Figure 10.6 is a pulse sequence diagram of a segmented cine gradient-echo pulse sequence that has been adapted by inclusion of MRE MEGs. The MEGs are synchronized with the initiation of the pulse sequence that is derived from a physiological trigger such as the ECG or pulse plethysmograph waveform [5]. Using the thin spherical shell model described above, Kolipaka et al. [5] has demonstrated that this pulse sequence (TE=9.8 ms; TR=35 ms; flip angle=30°; slice thickness=10 mm; $kx/ky=$ 256/64; FOV=14 cm; receiver bandwidth= 32 kHz; excitation frequency=200 Hz; four MRE time offsets; and 5 ms duration) has sufficient temporal resolution to be able to reproduce the variation in shear stiffness of a spherical shell phantom undergoing cyclic pressure variations

                                                                K.P. McGee et al.

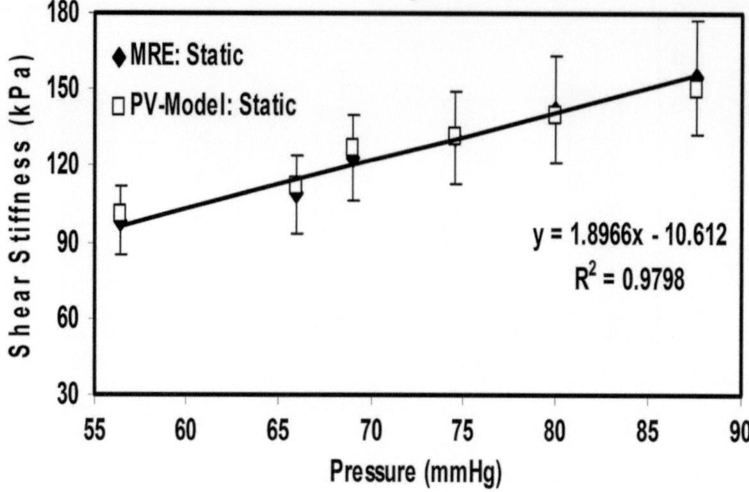

Fig. 10.4 MRE and PV-derived shear modulus as a function of static inflation pressure for a thin spherical shell

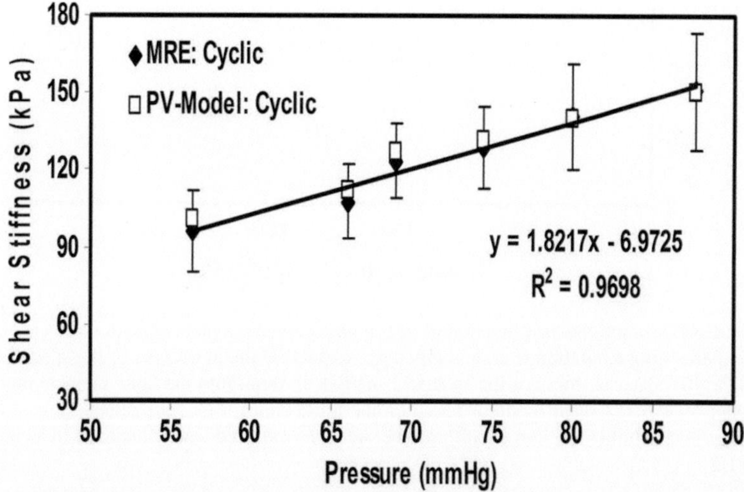

Fig. 10.5 MRE and PV-derived shear modulus estimates as a function inflation pressure measured at various points throughout a cyclic pressure cycle. The pressure represents the pressure within the cavity of the thin spherical shell which was generated by a computer controlled displacement pump connected to the thin spherical shell phantom

at a frequency of 1 Hz or 60 beats per minute or greater. The cine MRE pulse sequence showed excellent correlation between shear stiffness values obtained under the aforementioned dynamic conditions and those obtained when the phantom was inflated to same but static cavity pressures ($R^2 = 0.9948$).

In Vivo Validation of Cine Cardiac MRE

Recent research in cardiac MRE has focused on the evaluation of the previously described cine cardiac MRE pulse sequence and inversion algorithms in animal models as a precursor to clinical trials in humans. A study involving six live pigs demonstrated that MRE-derived estimates of shear modulus can be obtained in vivo and that these values track with pressure variations within the left ventricle of the heart throughout the cardiac cycle [6]. In this work, pressure values were obtained using a pressure catheter

inserted into the left ventricle. Figure 10.7 shows long and short axis views of the left ventricle of one animal obtained with and without application of cyclic motion from a pneumatic driver placed on the anterior chest of the animal above the heart. In the no motion case (Fig. 10.7a, e) the phase difference images which are used to encode cyclic displacements are dominated by noise and demonstrate the relative insensitivity of the ECG-gated GRE MRE pulse sequence to cardiac motion. When the MRE pulse sequence is synchronized to the pneumatic driver system, the complex wave pattern generated within the myocardium can be seen in both short and long axis views.

The inclusion of a pressure catheter into the ventricle of each animal allowed comparison of the MRE-derived shear stiffness estimates and pressure throughout the cardiac cycle. These results are shown in Fig. 10.8 and indicate that pressure and myocardial stiffness appear to track with one another throughout the cardiac cycle. Plotted as an ensemble of all six animals, the shear stiffness of the myocardium demonstrates a linear relationship between stiffness and cavity pressure as shown in Fig. 10.9.

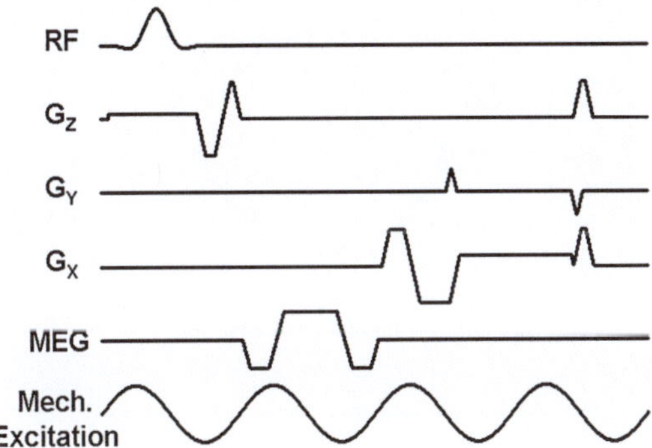

Fig. 10.6 Schematic of gradient echo MRE pulse sequence. The motion encoding gradients (MEGs) are designed to be gradient moment nulled and are synchronized to the mechanical excitation waveform. The figure also shows flow compensating gradients on X and Z axes respectively

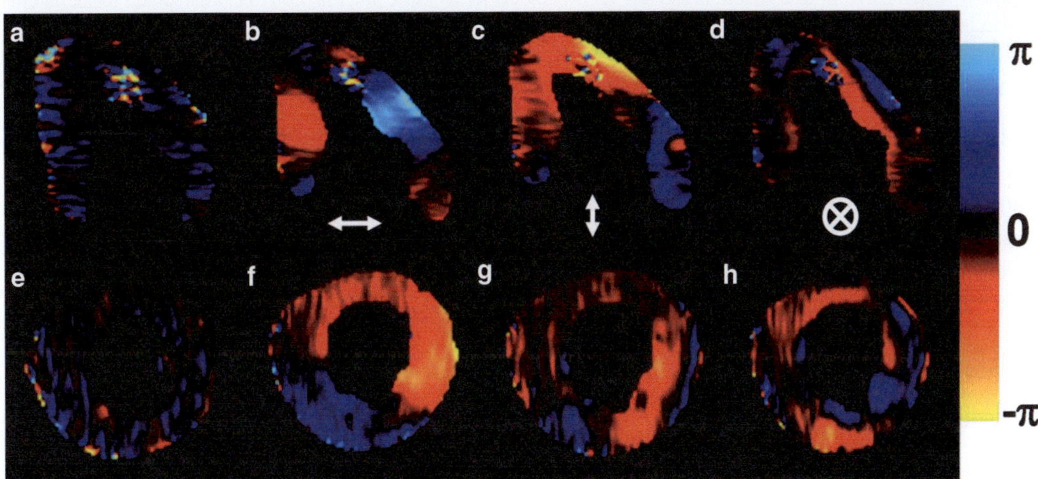

Fig. 10.7 End-diastolic short and long axis phase difference images of a porcine heart in vivo with and without the application of externally applied motion. In the absence of external motion, no shear waves can be seen (**a, e**). With the application of longitudinal excitation from two drum drivers located on the chest wall of the closed chest, shear wave induced displacements can be visualized within the two in-plane and through plane directions

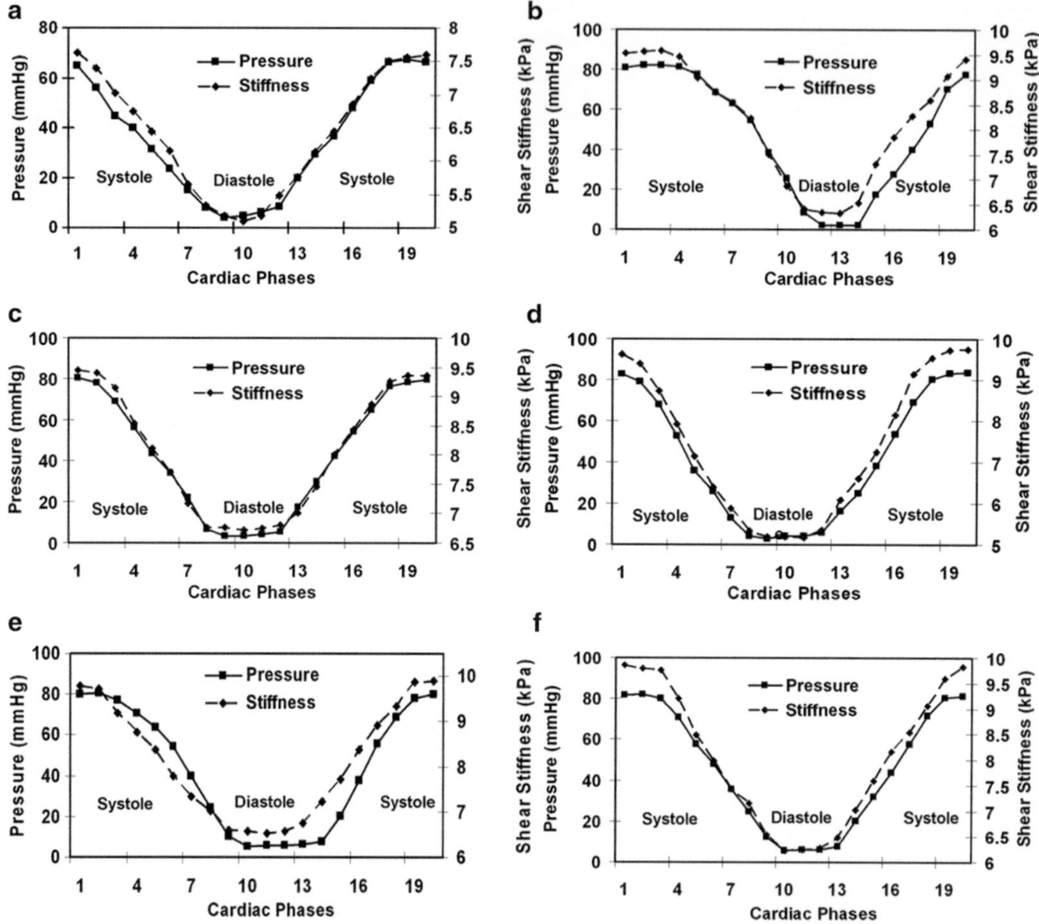

Fig. 10.8 MRE-derived shear modulus and left ventricular cavity pressure as a function of cardiac phase in six porcine in vivo hearts. Note that pressure and shear modulus values are displayed on separate scales

Future Directions

Future research will involve technical improvements to existing MRE image acquisitions and inversion algorithms as well as the application of cardiac MRE to a variety of cardiac diseases. For image acquisition, the immediate goal is to shorten scan times to allow the acquisition of volumetric data set covering the left ventricle. A complete data set would require obtaining displacement data in three encoding directions (e.g., physical x, y, and z), with four phase offsets, obtained at multiple points over the cardiac cycle. Acquiring all of this data in a single breath hold will require a combination of MRI techniques. However, once achieved this type of volumetric acquisition would allow the development of inversion algorithms incorporating displacements in all three directions. There are a number of potential clinical applications for cardiac MRE including heart failure with preserved ejection fraction and hypertrophic cardiomyopathy, both of which are thought to increase myocardial stiffness. While hypertrophic cardiomyopathy is currently diagnosed based on increased left ventricular wall thickness, there is no way to detect this disease before the wall thickens. In addition there is no way to distinguish hypertrophic cardiomyopathy from normal,

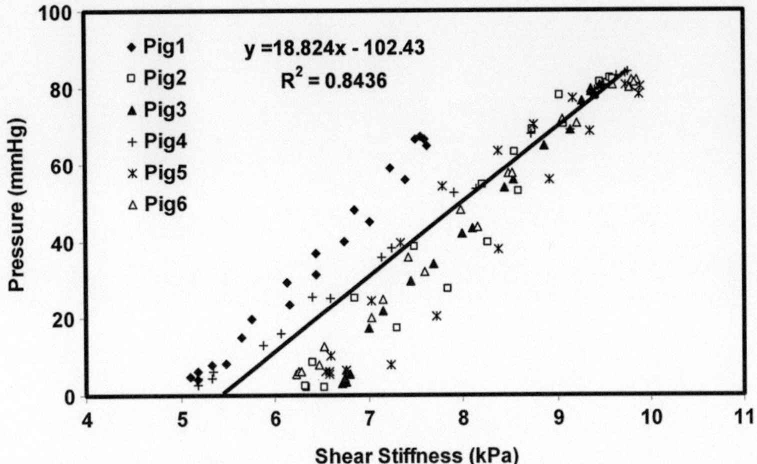

Fig. 10.9 Stiffness plotted as a function of left ventricular cavity pressure from six porcine hearts. Data was acquired in vivo throughout the animal's cardiac cycle. The data demonstrates an overall linear relationship between stiffness and pressure with a correlation coefficient of $R^2 = 0.84$

physiologic causes of ventricular wall thickening such as athletic training. Finally, increased myocardial stiffness could be predictive of poor outcome in a variety of diseases. For example, patients with hypertension sometimes progress to heart failure and that process is thought to be related to the interplay of pressure, volume, and the mechanical properties of the myocardium. Likewise, after myocardial infarction the non-infarcted ventricle undergoes changes in shape and function that are thought to be a response to the presence of the stiff myocardial infarct.

References

1. Burkhoff D, Mirsky I, Suga H. Assessment of systolic and diastolic ventricular properties via pressure–volume analysis: a guide for clinical, translational, and basic researchers. Am J Physiol Heart Circ Physiol. 2005;289(2):H501–12.

2. Mirsky I, Parmley WW. Assessment of passive elastic stiffness for isolated heart muscle and the intact heart. Circ Res. 1973;33(2):233–43.

3. Kolipaka A, McGee KP, Manduca A, Romano AJ, Glaser KJ, Araoz PA, Ehman RL. Magnetic resonance elastography: inversions in bounded media. Magn Reson Med. 2009;62(6):1533–42.

4. Kolipaka A, McGee KP, Araoz PA, Glaser KJ, Manduca A, Romano AJ, Ehman RL. MR elastography as a method for the assessment of myocardial stiffness: comparison with an established pressure–volume model in a left ventricular model of the heart. Magn Reson Med. 2009;62(1):135–40.

5. Kolipaka A, McGee KP, Araoz PA, Glaser KJ, Manduca A, Ehman RL. Evaluation of a rapid, multiphase MRE sequence in a heart-simulating phantom. Magn Reson Med. 2009;62(3):691–8.

6. Kolipaka A, Araoz PA, McGee KP, Manduca A, Ehman RL. Magnetic resonance elastography as a method for the assessment of myocardial stiffness throughout the cardiac cycle. Magn Reson Med. 2010;64(3):862–70.

Jun Chen, Jennifer Kugel, Meng Yin,
and Sudhakar Kundapur Venkatesh

MR Elastography of Breast

Introduction

Breast cancer ranks as the second leading cause of cancer deaths in women (after lung cancer). Early diagnosis of breast cancer and effective monitoring of treatment efficacy are critical for reducing breast cancer mortality. Mammography, ultrasound, and MRI play an important role in detection of breast cancer and have reduced breast cancer mortality. However, mammography may miss breast cancer especially in women with dense breast tissue, ultrasound has low sensitivity and specificity, and MRI has low specificity resulting in many false positives and unnecessary biopsies of benign tissues. Palpation wisdom from thousands of years has shown that breast cancer is usually much harder than normal breast tissue (like rock vs. marshmallow). Magnetic resonance elastography (MRE) is a nonionizing and noninvasive imaging modality that measures the stiffness of breast tissue and could potentially improve the specificity of breast MRI. This section will first briefly review the current breast cancer imaging techniques and emerging techniques, and then will focus on the MRE and discuss recent developments for breast cancer imaging.

Clinical Significance of Breast Cancer

A woman born in the United States today has a one in eight chance of having invasive breast cancer during her lifetime [1]. An estimated 39,970 breast cancer deaths (39,520 women, 450 men) were expected in 2011. Risk factors include age, certain gene traits, family history, early menarche, late menopause, postmenopausal obesity, use of combined estrogen and progestin menopausal hormones, alcohol consumption, and physical inactivity. Breast cancer typically has no symptoms when the cancer is in its early stages and most treatable. In the USA, considering all races, the 5-year relative survival is 99 % for localized breast disease when the cancer is confined to the breasts, 84 % for regional disease when tumors have spread to surrounding tissue or nearby lymph nodes, and 23 % for distant-stage disease when cancers have metastasized (spread) to distant organs. When breast cancer has grown

J. Chen, Ph.D. • M. Yin, Ph.D.
S.K. Venkatesh, M.D., F.R.C.R. (✉)
Department of Radiology, Mayo Clinic College of
Medicine, Mayo Clinic, 200 First Street SW,
Rochester, MN 55905, USA
e-mail: chen.jun@mayo.edu; yin.meng@mayo.edu;
venkatesh.sudhakar@mayo.edu

J. Kugel
Department of Radiology, Mayo Clinic College of
Medicine, Mayo Clinic, 200, First Street SW,
Rochester, MN 55905, USA

Diagnostic Radiology Research, Mayo Clinic,
Rochester, MN USA
e-mail: kugel.jennifer@mayo.edu

S.K. Venkatesh and R.L. Ehman (eds.), *Magnetic Resonance Elastography*,
DOI 10.1007/978-1-4939-1575-0_11, © Springer Science+Business Media New York 2014

to a size that can be felt, the most common physical sign is a painless lump. Sometimes breast cancer can spread to underarm lymph nodes and cause a lump or swelling, even before the original breast tumor is large enough to be felt. Less common signs and symptoms include breast pain or heaviness; persistent changes to the breast, such as swelling, thickening, or redness of the skin; and nipple abnormalities such as spontaneous discharge (especially if bloody), erosion, inversion, or tenderness [2].

Early diagnosis of breast cancer is a critical part of reducing breast cancer mortality because treatment is most effective at the early stages of disease. Death rates for breast cancer have steadily decreased in women since 1990, with larger decreases in younger women. From 2004 to 2008, rates decreased 3.1 % per year in women younger than 50 years of age and 2.1 % per year in women 50 and older. The decrease in breast cancer death rates represents progress in both earlier detection and improved treatment [2]. Right now, American Cancer Society guidelines for the early detection of breast cancer in average-risk, asymptomatic women are to have clinical breast examinations (CBE) at least every 3 years for women 20-39 years of age, and every year prior to mammography in women over age 40 [2].

Therefore, it is very important to optimize existing breast imaging techniques or to develop new techniques for accurate breast cancer diagnosis so that more breast cancers can be found and treated early, leading to the reduction of the breast cancer mortality.

This section will first briefly review the diagnosis accuracy of major conventional breast cancer imaging techniques and some emerging techniques, and then will focus on the MRE and discuss its new developments for breast cancer imaging.

Breast Cancer Imaging Techniques

Although mammography is the principle tool for diagnosing breast cancer, it is significantly less accurate in young women with mammographically dense breast tissue, who also have a higher risk of developing breast cancer than women with less dense breast tissue [3]. Also, mammography, together with other recently introduced ionizing radiation based techniques such as breast-specific gamma imaging (BSGI), positron emission mammography (PEM), digital breast tomosynthesis, and dedicated breast computed tomography, all have fatal radiation-induced cancer risks [4]. Another frequently used imaging modality, contrast-enhanced magnetic resonance imaging (CE-MRI) has a high sensitivity (89-100 %) for detecting breast cancer, but its specificity can be as low as 30 % [5]. Sonography is another major modality to detect breast cancer, however it has variable sensitivity (57–97 %) and specificity (60–90 %), which depends on many factors including the experience of the sonographer [6, 7]. The experience of readers also causes low interobserver agreement, for example, the κ values ($\kappa = 1$ means same reading) in Breast Imaging-Reporting and Data System (BI-RADS) category four were 0.33 (fair), 0.32 (fair), and 0.17 (poor) for the subdivisions 4a (low suspicion of malignancy), 4b (intermediate suspicion of malignancy), and 4c (moderate suspicion of malignancy), respectively [8].

Some emerging breast imaging techniques are still immature and need further development because their sensitivity and specificity varied greatly between different studies, such as Digital Infrared Thermal Imaging (DITI, Sensitivity = 0.25–0.97, Specificity = 0.12–0.85), Electrical Impedance Scanning (EIS, Sensitivity = 0.26–0.98, Specificity = 0.08–0.81), and ultrasound elastography (Sensitivity = 0.35–1.00, Specificity = 0.21–0.99) [9]. In the past few years, ultrasound elastography techniques, categorized as qualitative (strain) and quantitative (stiffness) methods, have been investigated widely and have shown potential for improving the accuracy of breast cancer diagnosis alone or in conjunction to B mode Sonography [10–13]. However ultrasound elastography techniques are highly operator-dependent, and suffer from low image quality (49.6 % cases) which can significantly decrease the sensitivity to as low as 56.8 % [14]. Qualitative ultrasound elastography (strain imaging) also has trouble with characterizing benign breast lesions such as fibrocystic changes and fibroadenomas due to their low contrast to surrounding normal breast tissues, while quantitative ultrasound elastography (stiffness imaging) has difficulties with imaging the tissue deep in the breasts, especially for malignant tumors [11].

MR Elastography of Breast

In a clinical breast examination or breast self-examination, the cancer lumps can be felt by palpation if they are large enough. The lumps are often "firmer" or "stiffer" than normal breast tissues. In biomechanical terms, "firmness" or "stiffness" corresponds to defined biomechanical properties of tissues. Like the difference we can feel between "firm" and "soft" mattresses, or ripe and unripe fruit, we can also feel the stiffness difference between firm tissue (like muscle or a solid cancerous lumps) and soft tissue (like fat) using palpation. Malignant tumors often infiltrate into the surrounding tissue and lead to a pronounced reactive proliferation of connective tissues, which increases tissue stiffness [15]. Krouskop et al determined that the contrast between cancerous and noncancerous breast lesions was 1,000–5,000 % [16]. However, palpation is subjective, qualitative (cannot measure the actual tissue stiffness values) and may not be sensitive enough to detect small lumps (<10 mm) or lumps that are too deep in the breasts.

To accurately and objectively measure tissue stiffness, even for small lesions deep in the breasts, MRE techniques have been developed over the last 20 years. The basic principles of MRE of the breast are as follows: first, an external mechanical driver is used to apply alternate/quasi-static forces to the breasts causing the internal mechanical motion (shear wave) within the breasts; second, a motion sensitive MRI sequence is used to image and measure the internal mechanical motion; last, tissue stiffness is computed because the mechanical motion is a function of tissue stiffness [17]. In fact, ultrasound elastography follows the similar steps, except it uses ultrasound to measure the motion in tissue and uses the ultrasound transducer (or remote

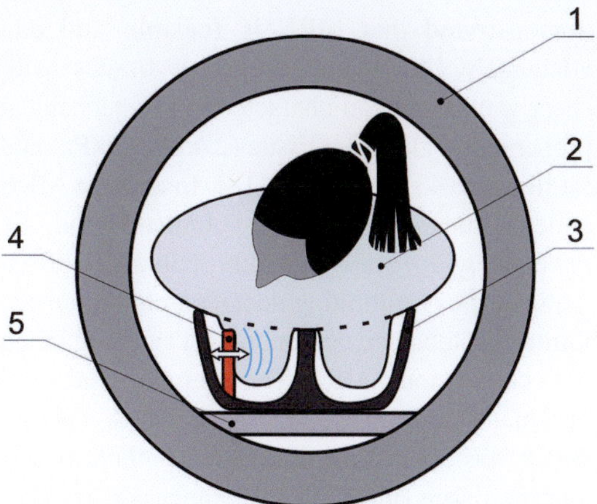

Fig. 11.1 A typical conventional compressive breast MRE setup. 1—MRI scanner bore, 2—patient in prone position, 3—breast RF coil, 4—compressive MRE driver, 5—patient table. The compressive breast MRE driver requires contact and compression of the breast tissue to maintain a good mechanical coupling for transmitting acoustic waves into breast tissues. The driver can be positioned on the *right, left,* superior, inferior, or/and anterior part of one or both breasts. If both breasts are contacting the driver, then they can be imaged by MRE at the same time

radiation force) to push against the breasts for exciting the motion inside the breast tissue. Compared to ultrasound elastography, MRE has overall better image quality, which leads to higher accuracy of measuring breast tissue stiffness.

Figure 11.1 shows the conventional setup of breast MRE, which often involves five steps before the MRI/MRE scan. (1) External MRE drivers are positioned inside one or two cavities of a breast RF coil. The drivers should be in contact with one or both breasts, either on the lateral aspect or anterior aspect of the breast. (2) The subject lies on top of the RF coil in a prone position with their breasts hanging inside the cavities, similar to other routine breast MRI exams. (3) Depending on the size of breasts, the operator usually needs to adjust the external drivers so that the drivers are in good contact with the breasts for transmitting acoustic waves to the breasts, which often involve a precompression to the breasts. (4) Reposition of the drivers may be required if there is any discomfort due to driver setup. (5) The patient is then ready for MRI and MRE scans.

In one of the first in-vivo breast MRE studies, nine healthy female volunteers were studied and demonstrated that MRE is feasible and can adequately illuminate the breast tissues with shear waves and can characterize biomechanical properties of glandular tissue (2.45 ± 0.2 kPa) and fat tissue (0.43 ± 0.07 kPa) [18]. In another MRE patient study, six healthy volunteers and six patients with biopsy-proven palpable breast malignancies (infiltrating ductal carcinoma, $n=5$; infiltrating lobular carcinoma, $n=1$) were scanned with conventional MRE. The average shear stiffness of the tumors was 33 kPa (range = 18–94 kPa), which was about four times greater than that of adipose tissue (mean = 8 kPa, range = 4–16 kPa) in breast cancer patients. In the healthy volunteers, the mean value for adipose tissue was 3.3 ± 1.9 kPa, which is less than their fibroglandular tissue (7.5 ± 3.6 kPa) [19].

In addition to the shear modulus of breast tissue, Sinkus et al. [20, 21] have extended MRE to measure the eigenvalues of elasticity tensor of breast tissues in patients with breast carcinoma and found that carcinoma exhibits an anisotropic elasticity distribution while the surrounding benign tissue appears isotropic. It is well known to physicians that some malignant lesions feel stiff while others feel soft. Additional information about the anisotropy of the lesion has the potential to increase the diagnostic accuracy of MRE in the detection of breast malignancy [20].

Xydeas et al. [15] studied viscosity and elasticity of breast tissues in 5 patients with 6 malignant lesions, 11 patients with benign lesions, and 4 patients with no lesions using MRE. The mean viscoelastic parameters were: breast cancer (elasticity = 3.1 ± 0.7 kPa, viscosity = 2.1 ± 1.2 Pa∗s), fibroadenoma (elasticity = 1.4 ± 0.5 kPa, viscosity = 1.7 ± 0.8 Pa∗s), fibrocystic changes (elasticity = 1.7 ± 0.8 kPa, viscosity = 1.7 ± 0.9 Pa∗s), and surrounding tissue (elasticity = 1.2 ± 0.2 kPa, viscosity = 1.0 ± 0.31 Pa∗s). In terms of elasticity values, malignant lesions were significantly stiffer than fibroadenoma ($P < 0.0004$), fibrocystic changes ($P < 0.04$), and overall breast surrounding tissues ($P < 0.0005$), while for viscosity values, there were no significant differences between them. Two patients with fibroadenoma had a dynamic breast MR imaging score (in accordance with the BI-RADS system) of 3 out of 4, and small elasticity values. MRE correctly identified these as benign lesions which would have been diagnosed as malignant tumors based on contrast-enhanced MRI features.

It is known that tissue elasticity and viscosity are reflected by directly measured complex shear modulus, $E = Er + i*Ei$, with Er and Ei being the real and imaginary part respectively. However, complex values are not straightforward to understand in practice; therefore, different rheological models (Voigt model, Maxwell model, and spring pot model) have been tested to explain the breast tissue biomechanical behavior, but none of them fits perfectly [22]. Therefore, Sinkus et al. followed a pure physically motivated approach, power law model, to study the viscoelastic behavior of breast lesions in the perspectives of solid/liquid duality ($y=0$ for pure solid, $y=0.5$ for pure liquid) and attenuation (α_0, at 1 Hz) [22]. This study included 68 female patients with breast tumors (benign: $n=29$, malignant: $n=39$), and revealed that malignant lesions tend to have high y (more liquid-like) values and low α_0 values (less attenuation), which is opposite to the behavior of benign lesions. The interpretation of complex shear modulus in terms of y and α_0 now provides an understanding of the underlying physics: malignant lesions are characterized by an enhanced liquid-like (greater y values) behavior and a decreased scale of attenuation (small α_0 values) [22]. When y approaches to 0.5 (pure liquid), the power law model suggests a loosely connected network formed in tissue. This is consistent with the macroscopic alterations such as stromal reaction induced by the tumor, which leads to the remodeling of the normal extracellular matrix and changes in cell density. Angiogenesis, tumor invasion, and metastasis all require a degradation of the extracellular matrix [22]. Finally, when combined with contrast-enhanced MR mammography, the study showed that MRE significantly increased the specificity by 20 % (from about 40 % to about 60 %) at 100 % sensitivity for diagnosing malignancy (AUROC = 0.96 ± 0.02) [22]. In another similar study with 57 patients, MRE increased the specificity of CE-MRI from 75 to 90 % and sensitivity of 90 % and combined AUROC of 0.96 [23].

New Developments of Breast MR Elastography: Noncompressive 3D Technique

With the conventional MRE setup, all breast drivers contact and compress the breasts for efficient mechanical wave transmission. This can cause significant biomechanical property change because of the nonlinearity of biological tissues [21]. The compression is difficult to control and measure in practice, which leads to the question of the impact of compression on the accuracy of breast elastography methods, including ultrasound elastography and MRE. A recent human study has shown that for adipose and fibroglandular tissues, 10 % tissue compression approximately doubled the observed shear wave speed, which represents about a four-fold increase in stiffness. Adipose tissue was observed to increase in stiffness with the amount of compression; so did breast cancers at a much slower rate. The stiffness difference between adipose tissue and malignant lesions became smaller as the compression level increased, and eventually disappeared when the compression level was about 40 % [24]. This compression effect partly explains the inconsistencies in the reported stiffness values of breast tissues from different elastography methods in the literature [8, 10, 19, 23, 24].

A novel noncompressive breast MRE technique has been developed to eliminate the effect of driver compression on tissue stiffness change [25]. Figure 11.2 shows the diagram of the noncompressive breast MRE setup. Compared with the conventional compressive breast MRE setup, the noncompressive setup has the following advantages. (1) does not add tension or change the shape of the breasts, which avoids the biomechanical property change and geometry/boundary condition change; (2) the driver-breast mechanical coupling is not affected by the size of the breasts; (3) The breast RF coils does not require any modification or customization to accommodate the driver and is automatically compatible to CE-MRI setup, thereby promoting their use in clinical practice; (4) the driver does not interfere with MRI-guided breast biopsy, and (5) both breasts can be imaged by MRE at the

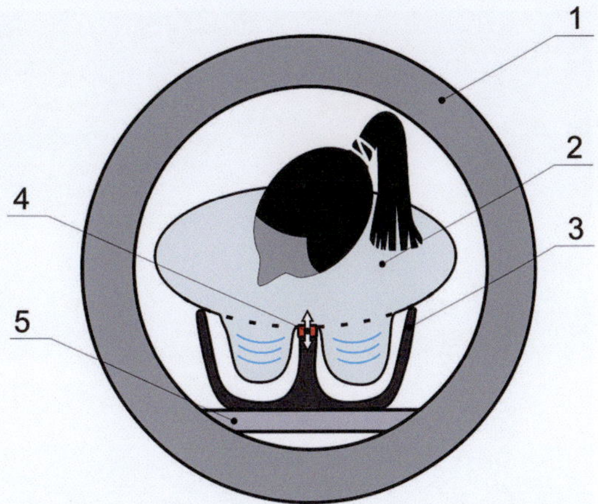

Fig. 11.2 A novel noncompressive breast MRE setup. 1—MRI scanner bore, 2—patient in prone position, 3—breast RF coil, 4—noncompressive MRE driver, 5—patient table. Instead of compressing breast tissue, the noncompressive breast MRE driver is only in contact with the patients' sternum to transmit acoustic waves into breast tissues through the chest wall. The driver is automatically compatible to CE-MRI setup. Both breasts can be imaged by MRE at the same time

same time. A typical 3D GRE MRE imaging sequence and inversions used in the noncompressive breast MRE are detailed in [26], with the following parameters. A 1.5-T MRI scanner (GE, Signa, Wisconsin, USA) and a MRE active driver system (Resoundant, Rochester, MN) were used. Imaging parameters: vibration frequency = 40 Hz; vibration power level = 20 %; $FOV_{x/y/z}$ = 30-34/30-34/14.4-18 cm; four phase offsets; motion-encoding gradient (MEG) amplitude = 2.8 G/cm; TR = 31.3 ms; TE = 27.2 ms (fat/water in-phase echo time); flip angle = 15°; BW = 31.25 kHz; axial imaging plane covering both breasts in the SI direction; acquisition matrix = 96×96×40; reconstruction matrix = 256×256×36; NEX = 1; SENSE acceleration factor = 2 (RL direction); total scan time = 9′54″ (free breathing). Inversion parameters: the vector curl of the measured wave data was calculated using 3×3×3 derivative kernels on the wrapped phase data acquired in three orthogonal directions [27]. A 3D local frequency estimation (LFE) inversion was performed on the curl data with 2D directional filters (cut-off frequencies of 2 and 128 cycles/FOV) to calculate the volumetric elastograms of the two breasts [28].

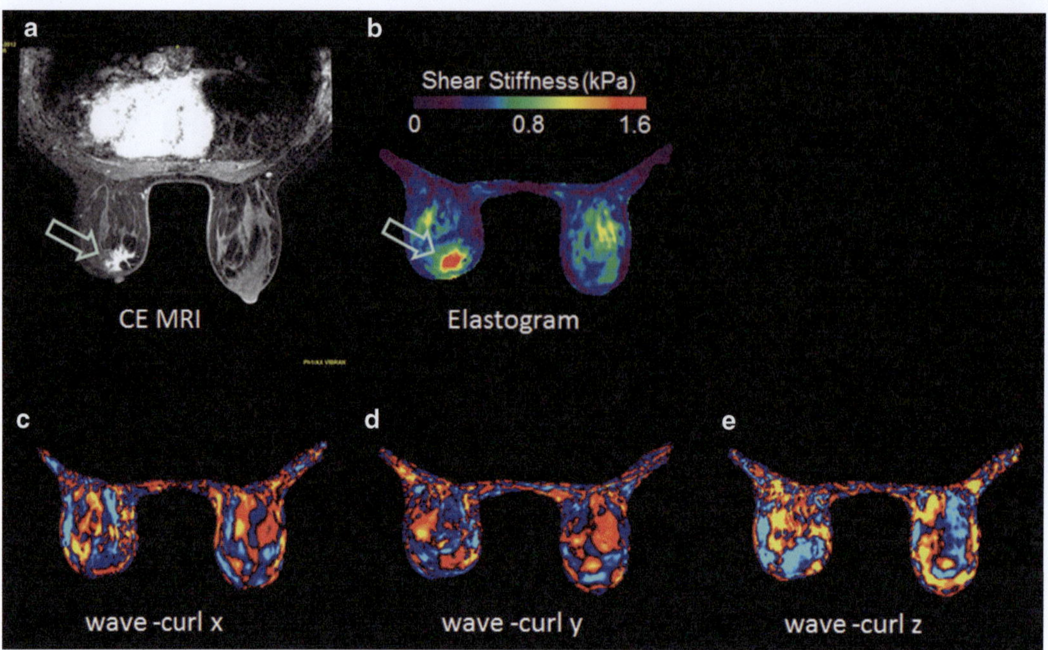

Fig. 11.3 Example MRE images of a 41-year-old-female patients with invasive ductal carcinoma. (**a**): CE-MRI; (**b**): MR elastogram; (**c**), (**d**), and (**e**): x, y, and z component of curl displacement respectively

A preliminary study was performed with seven healthy normal female volunteers without known breast disease and a 41-year-old female patient with a biopsy-proven invasive ductal carcinoma. Bilateral noncompressive MRE with the above parameters was performed in all subjects, and the patient also underwent a CE-MRI study. In Fig. 11.3, CE-MRI shows that in the left breast of this patient, there is a heterogeneously enhancing mass in the right subareolar breast tissue corresponding to the biopsy-proven malignancy, with a size of $3.2 \times 2.0 \times 2.4$ cm (Fig. 11.3a, arrow). No abnormal indications were seen in the right breast. The breast MR elastogram shows that the glandular tissue is heterogeneous in stiffness, and the carcinoma is much stiffer than the surrounding breast tissue (Fig. 11.3b, arrow). The x, y, and z curl components of the measured wave displacement show adequate wave propagation in both breasts (Fig. 11.3c, d, e). For this patient, the stiffness of adipose tissue was 0.41 ± 0.10 kPa, glandular tissue was 0.90 ± 0.18 kPa, and the invasive ductal carcinoma was 1.42 ± 0.17 kPa. In the seven normal volunteers, the stiffness of adipose tissue ranged from 0.25 to 0.41 (mean$=0.33$) kPa and glandular tissue ranged from 0.46 to 0.9 (mean$=0.64$) kPa [29]. The invasive ductal carcinoma is about 3 times stiffer than the adipose tissue and 1.5 times stiffer than the glandular tissue.

Conclusions

Breast cancer can be treated more effectively in early stages than in late stages, and accurate diagnosis diagnostic methods are desired to detect breast cancer early. MRE is a promising for differentiating malignant from benign breast lesions. A recently developed noncompressive MRE technique can complement the CE-MRE setup, can image both breasts at the same time, and applies no compression to the breasts, which avoids the tissue biomechanical property change due to its nonlinearity. Future work will include recruiting a larger number of breast cancer patients to evaluate the noncompressive MRE to diagnose breast cancer as a single technique and in conjunction with CE-MRI for improving its specificity.

MRE of the Spleen

The spleen is situated in the left upper abdomen below the diaphragm and performs the function of hemopoiesis and acts as a reservoir of blood in acute stress. The spleen is intimately related to the portal circulation and splenomegaly in a patient with chronic liver disease and cirrhosis is suggestive of portal hypertension. Therefore, spleen has been evaluated both clinically and by imaging for an increase in size as an evidence for portal hypertension. However there is poor correlation between spleen size and severity of portal hypertension [30] and splenomegaly may be related to tissue hyperplasia and fibrosis in addition to congestion [31]. Evaluation of splenic stiffness may provide further insights into the pathophysiology of splenomegaly and the complex relationship between liver fibrosis, portal hypertension, and spleen. Several studies have focused on the evaluation of spleen stiffness with MRE.

MRE of the spleen can be performed by placing a passive driver over the spleen [32, 33]. Custom made pillow-like passive actuator flexible drivers (Fig. 11.4) that are made of a soft inelastic fabric cover around a porous, springy mesh core [34] can be easily placed on the curved lateral abdominal wall. In a preliminary study [33], a significant difference in measured splenic stiffness was found when performed with the driver placed over the liver on right abdomen and with the driver over the spleen on left side of the abdomen. MRE of the spleen is best performed with the driver over the left side (over the spleen) to ensure through plane propagation of shear waves in the spleen.

The mean stiffness of normal spleen tissue measured in healthy volunteers ranges from 2.35 to 5.6 kPa with an average value of 3.6 kPa. No significant correlation was demonstrated between splenic stiffness and age, sex, body mass index and arterial mean blood pressure [32, 33].

Evaluation of splenic stiffness in chronic liver disease and portal hypertension has been the subject of interest in many studies including those with transient elastography. In a large animal model, Nedredal et al. showed significant correlation between spleen stiffness with MRE and measured direct portal vein pressure gradient [35]. In another well-controlled in-vivo porcine model study, spleen stiffness and liver stiffness were shown to correlate with measured portal pressure [36].

In a preliminary human study with 38 patients with chronic liver disease, Talwalkar et al. showed significant correlation between liver stiffness and spleen stiffness ($r^2 = 0.75$; $p < 0.001$) and a systematic increase in spleen stiffness with increasing liver fibrosis stages (Fig. 11.5). A mean spleen stiffness of ≥ 10.5 kPa was associated with esophageal varices (Fig. 11.6). In another study, Morisaka et al. [37] demonstrated

Fig. 11.4 A ergonomic flexible pillow passive driver. The flexible driver can be easily applied to curved surfaces like abdomen and chest walls which ensure good contact with the surface and better transmission of shear waves

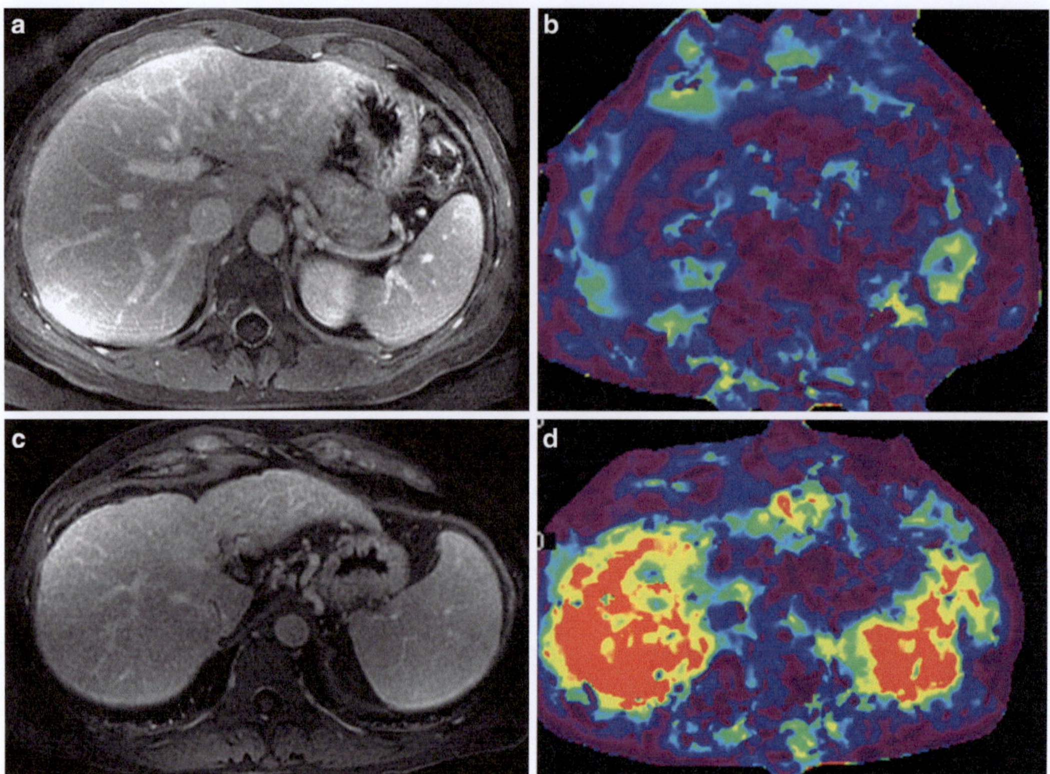

Fig. 11.5 Splenic stiffness in chronic liver disease. Post contrast enhanced images (**a,c**) and stiffness maps (**b,d**). An example of simple fatty liver only (*top row*) with normal liver stiffness of 1.9 kPa and spleen stiffness of 3.2 kPa. A case of chronic hepatitis C with cirrhosis (*bottom row*) shows elevated liver stiffness of 8.2 kPa and that of spleen to 9.1 kPa consistent with portal hypertension

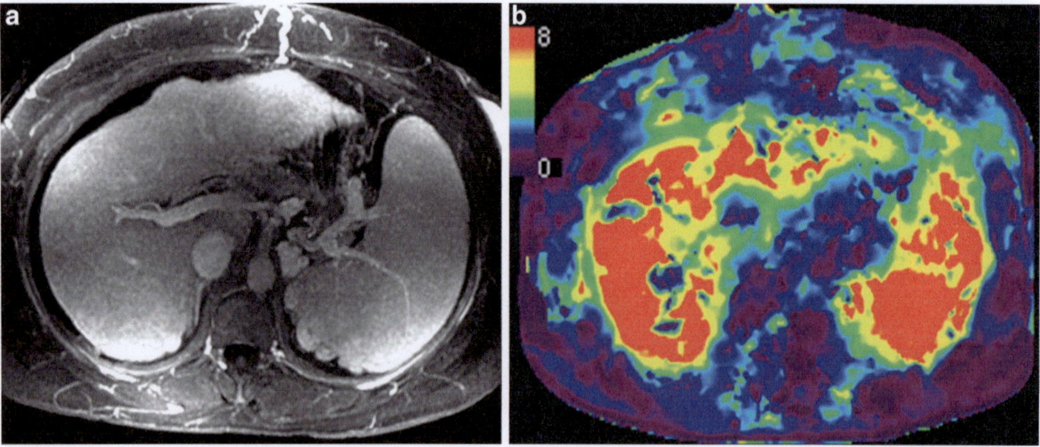

Fig. 11.6 Splenic stiffness can predict esophageal varices. An example of NASH with cirrhosis and portal hypertension. Contrast-enhanced MR image (**a**) showing nodular liver, splenomegaly, splenic and gastric collaterals, and dilated anterior abdominal veins. Stiffness map (**b**) shows severely elevated splenic stiffness to 11 kPa and that of the liver to 10.5 kPa. Endoscopy confirmed esophageal varices

strong association of splenic stiffness with severe gastroesophageal varices with a cut-off value of 10.1 kPa similar to the Talwalkar study.

Shin et al. [38] using a 3D echo planar MRE technique showed that both hepatic and splenic stiffness had positive linear correlations with the endoscopic grade of esophageal varices. The diagnostic performance of hepatic and spleen stiffness in predicting high-risk varices was comparable to that of dynamic contrast-enhanced (DCE) MR imaging and DCE+MRE had higher sensitivity in predicting esophageal varices and high-risk varices. In a recent study, Ronot et al. [39] showed significant correlation of hepatic venous pressure gradient (HVPG), with liver and spleen loss modulus assessed with MRE. Severe portal hypertension and high-risk varices were best identified with spleen loss modulus.

Assessment of splenic stiffness is useful in the evaluation of portal hypertension for the prediction of esophageal varices and high-risk varices. In the future, spleen stiffness may also serve as a quantitative parameter for the assessment of response to treatment of portal hypertension.

MRE of the Pancreas

The pancreas is a midline retroperitoneal organ situated in a roughly transverse orientation in the upper abdomen. The location of the pancreas makes propagation of shear waves challenging. However it is possible to introduce propagating shear waves by placing a modified large driver in the midline of the upper abdomen [40]. Due to the small size of the gland and complex shape, a three-dimensional (3D) analysis of the wave field would be required. Ideally the patient should be fasting to ensure an empty stomach so that distance between the driver and pancreas is reduced as much as possible to ensure good propagation of waves. This is also reduces possible compression by a full stomach.

In an initial study of healthy volunteers by Yin et al. [41], the mean shear stiffness of the pancreas was 2.0±0.4 kPa at 60 Hz. Recently Yu et al. [40] showed that MRE was feasible in normal volunteers using an ergonomic soft pillow driver to deliver vibrations to the pancreas. In this study with 20 healthy volunteers, the propagating shear waves were imaged with a 2D multi-slice echo planar imaging based MRE pulse sequence and at 40 and 60 Hz. The mean stiffness of pancreas was 2.09±0.33 kPa at 60 Hz and similar to liver, whereas at 40 Hz, it was 1.15±0.17 kPa, significantly lower than the liver. The stiffness in different regions of the pancreas was similar with no significant differences (Fig. 11.7). The initial studies have provided motivation for application of MRE in evaluation of pancreatic diseases. MRE of the pancreas may be useful in the assessment of chronic pancreatitis as well as focal lesions in the pancreas such as adenocarcinoma.

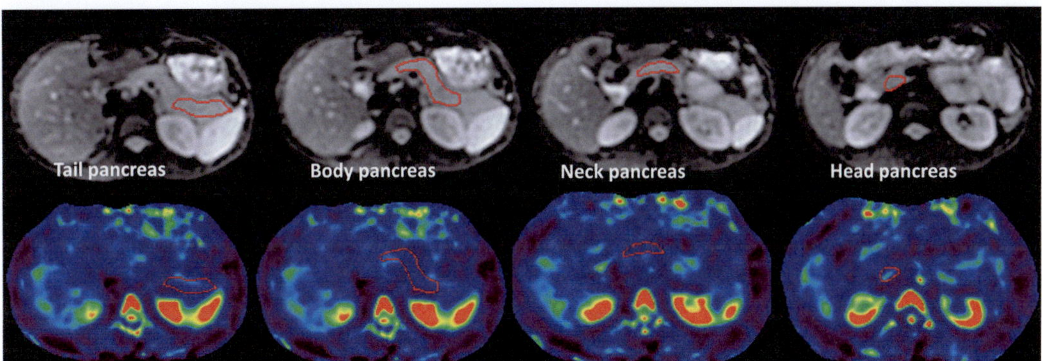

Fig. 11.7 MRE of the pancreas. Anatomical MR images (*top row*) and stiffness maps (*bottom row*) through different regions of pancreas in a normal healthy volunteer. A manual region of interest has been drawn in different parts for measurement of stiffness (*red outlined areas*). The mean stiffness of the pancreas was 2.08±0.3 kPa (color figure online)

MRE of the Kidney

Kidneys are bilateral small bean-shaped organs in the retroperitoneal region. The location and small size of the kidneys pose challenges for introducing shear waves from an external driver similar to pancreas. In a preliminary study, Venkatesh et al. showed the feasibility of MRE of the kidneys in healthy volunteers [42] with the subject lying supine and on top of a single large disc-shaped driver at the level of kidneys placed in contact with the posterior abdominal wall. In another study Rouviere et al. showed that MRE of the kidney was feasible and reproducible with an intra-subject variability of 6 % [43]. In this study, they studied the left kidney separately with a drum-like passive driver placed at the level of the kidney in contact with the posterior abdominal wall. In these preliminary studies, MRE was performed at different frequencies ranging from 45 to 90 Hz. MRE of the kidneys can also be performed with two drivers positioned at the level of the kidneys and in contact with the posterior abdominal wall (Fig. 11.8). In view of the small size of the kidneys, a higher frequency (90 Hz) and a 3D MRE technique would be ideal. MRE obtained in the coronal plane including both kidneys would be useful for comparison (Fig. 11.9).

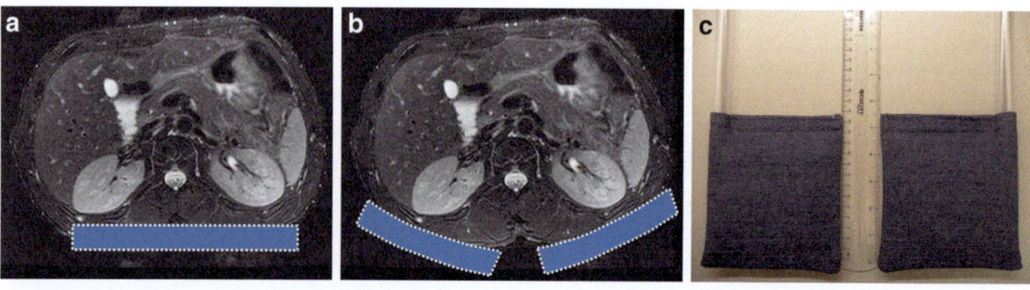

Fig. 11.8 MRE of the kidneys. MRE of the kidneys can be performed using a large driver (blue box) placed in the midline (**a**) or with two small drivers (blue boxes) placed on either side of the spine (**b**). Ergonomic flexible soft pillow drivers for MRE of the kidneys (**c**)

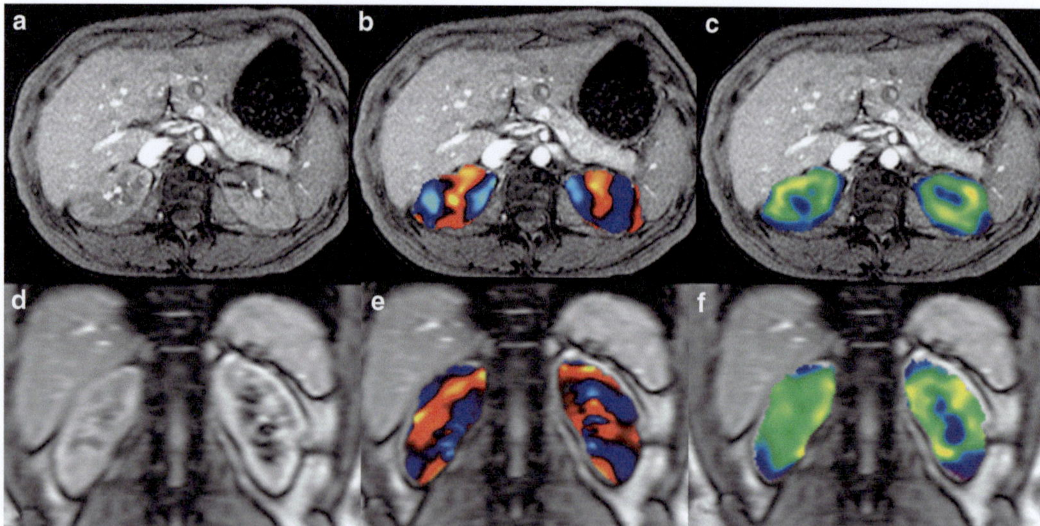

Fig. 11.9 MRE of the kidneys in a normal healthy volunteer performed with a single large driver placed in the midline and in contact with the posterior abdominal wall. MRE in an axial plane obtained at 60 Hz (*top row*) and in a coronal plane obtained at 90 Hz (*bottom row*). Anatomical MR images (**a, d**), superimposed wave images (**b, e**), and stiffness maps (**c, f**) on the anatomical images. Note excellent transmission of shear waves in the kidneys at both frequencies

Renal tissue stiffness is dependent on several components of the parenchyma. The kidneys receive nearly 25 % of the cardiac output and up to 25 % of renal volume may be attributable to blood pressure and the content of blood, filtrate, and urine in the kidney [44]. Changes in renal blood flow have been shown to affect renal stiffness [45], independent of the presence of fibrosis. Therefore renal perfusion needs to be considered when interpreting renal stiffness.

MRE for the evaluation of renal parenchymal disease has been studied in both animals and humans including renal allografts [42, 43, 45–49]. These studies have demonstrated feasibility and have also shown regional differences in the viscoelastic properties. A recent study [50] has shown that higher resolution MRE of native kidneys is possible and renal medulla has higher stiffness than cortex.

MRE of Renal Allograft

MRE of the renal allograft is probably easier to perform as the graft is usually located in the lower abdomen where it is subject to less movement due to respiration, and near the skin surface which would facilitate transmission of shear waves. Also MRE with a higher frequency would be easier due to less attenuation of the high frequency shear wave as the graft is superficial in location (Fig. 11.10).

In recent years, evaluation of renal allografts for interstitial fibrosis has generated interest among several investigators. Interstitial fibrosis and tubular atrophy (IFTA) is the most common cause of renal allograft failure. It is hypothesized that the stiffness of allografts with IFTA will be higher as there is diffuse interstitial fibrosis, which may increase tissue stiffness similar to increased stiffness in liver fibrosis and cirrhosis. Studies with MRE in renal allografts are few. In one study, Lee et al. [49] showed a trend of increased stiffness in patients with mild and moderate fibrosis. However no significant differences were found in this study. The study was comprised of only 11 patients with interstitial fibrosis (mild 7, moderate 2, and not significant in 1). Another study by Venkatesh et al. [51] demonstrated higher stiffness in patients without IFTA (2 patients) as compared to those with IFTA (16 patients). This study was comprised of ten patients with mild, one patient with moderate, and two patients with severe IFTA. Both these studies demonstrate complex relationships between renal stiffness and interstitial fibrosis. The exact reasons are not understood well but may be explained by the loss of the micro capillary network that usually accompanies or precedes IFTA. This leads to reduced perfusion that may reduce tissue stiffness. Future studies are needed to tease out the relationship between renal allograft stiffness and interstitial fibrosis and renal perfusion.

Clearly MRE of the kidneys is an exciting field with possible clinical applications in the evaluation of chronic parenchymal disease characterized by renal fibrosis and IFTA in renal allografts.

MRE of the Uterus

The uterus is a hollow thick-walled muscular organ situated in the pelvis. The bulk of the uterus is made of a smooth muscle layer known as myometrium enclosed by a thin external covering

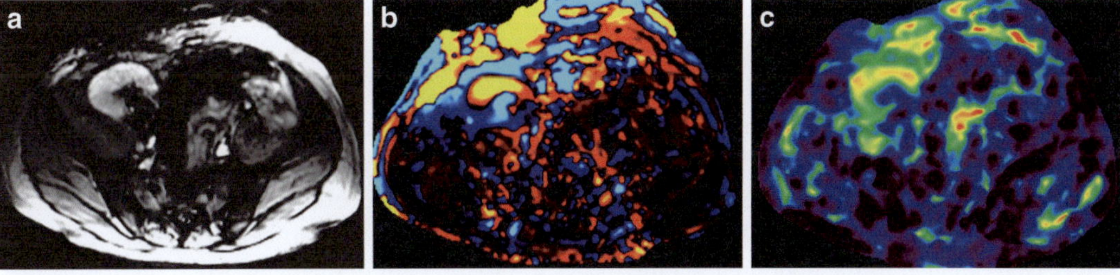

Fig. 11.10 MRE of a renal allograft performed at 90 Hz. Anatomical MR image (**a**), wave image (**b**), and stiffness map (**c**) of a normal renal allograft

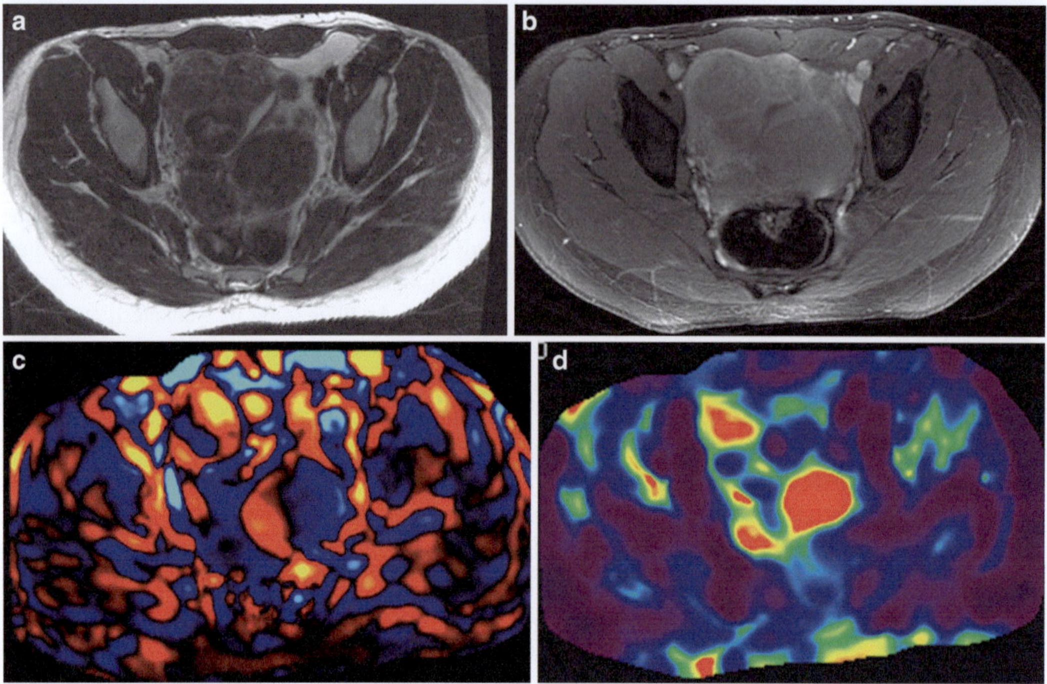

Fig. 11.11 MRE of uterus. MRE performed at 60 Hz in a patient with multiple fibroids. Axial T2 image (**a**), contrast-enhanced T1 image (**b**) of pelvis showing multiple fibroids in enlarged uterus. MRE wave image (**c**) showing good propagation of the shear waves through the uterus and stiffness map (**d**) showing stiff regions corresponding to the fibroids

known as perimetrium. Leiomyoma (fibroid) are the most common uterine neoplasms [52] and can cause menorrhagia, infertility, pregnancy loss, dysfunctional labor and pelvic pain. Fibroids arise from myometrium consisting of homogenous tissue of smooth muscle but enclosed in a capsule [53]. Fibroids are typically firmer or harder than surrounding myometrium and this is probably due to excessive extracellular matrix [54]. MRE may be useful in demonstrating elevated stiffness in the region of the fibroid and variability due to differences in composition.

In a preliminary study, Stewart et al. [55] showed feasibility of MRE for the evaluation of uterine fibroids. In this study, MRE of the uterus was performed with the patient lying supine on the MRI table. A circular pneumatic passive driver was placed on their lower abdomen above the uterus. This driver was affixed snugly using an elastic "belt." Continuous 60 Hz longitudinal waves were applied to the abdomen via the passive driver. A modified gradient-recalled echo (GRE) MRE sequence was used for the acquisi-

tion using the following parameters: 30° flip angle, 50 ms TR, 8 phase offsets through time, and all three motion sensitizing directions were acquired. A 4-mm thick transverse slice was prescribed through the center of the fibroid. The waves were analyzed using two criteria; depth of penetration and wave amplitude. The wave images were inspected to ensure that the waves penetrated the entire fibroid. Without wave penetration, elastography cannot be performed. If the waves penetrated the entire fibroid, they were analyzed to make sure that they had sufficient wave amplitude. The wave images were processed using a 2D local frequency estimation (LFE) algorithm to create a qualitative map of stiffness, called an elastogram. Manually drawn regions of interest (ROI) were placed on the resulting elastograms and the mean and standard deviation are recorded. The leiomyoma studied in the study by Stewart et al. ranged in size from 4.5 to 22.5 cm in diameter. MRE was performed successfully in all six patients. The stiffness ranged from 3.95 to 6.68 kPa with an average stiffness of 5.09 kPa. The study results showed that uterine fibroids have variable stiffnesses probably reflective of their tissue composition.

In another study, MRE was performed before and after ultrasound ablation (Exablate) of the fibroids. In this study, Hesley et al. used MRE to characterize the change in fibroids after treatment with MR-guided focused ultrasound ablation. It was shown that the fibroids become stiffer after ablation with focused ultrasound [56].

Studies have shown the feasibility of MRE of the uterine fibroid (Fig. 11.11), but more data are needed to further validate the initial findings. Three-dimensional MRE imaging will be needed to allow for the imaging of smaller fibroids.

MRE of Other Organs

Research studies have shown the feasibility of performing MRE for evaluation of aortic wall stiffness [57, 58], thyroid gland [59], tongue [60], and prostate [61–63] among others. Clinical applications may emerge in the future for a possible role of MRE in evaluation of diseases affecting these organs.

MR Elastography has proven to be a robust and versatile noninvasive technique for assessment of the stiffness of different organs. Continuing technical improvements set the stage for MRE in the evaluation of several organs and their pathologies that were previously inaccessible.

Acknowledgment We thank Kathleen R. Brandt, M.D. and Karthik Ghosh, M.D. for their expertise and advice.

References

1. Warner E. Clinical practice: breast-cancer screening. N Engl J Med. 2011;365(11):1025–32.
2. ACS. Breast Cancer Facts & Figures 2011–2012. Atlanta: American Cancer Society, Inc. 2012. Report No.
3. Hooley RJ, Greenberg KL, Stackhouse RM, Geisel JL, Butler RS, Philpotts LE. Screening US in patients with mammographically dense breasts: initial experience with Connecticut public act 09-41. Radiology. 2012;265:59–69.
4. Hendrick RE. Radiation doses and cancer risks from breast imaging studies. Radiology. 2010;257(1):246–53.
5. Kuhl C. The current status of breast MR imaging part I: choice of technique, image interpretation, diagnostic accuracy, and transfer to clinical practice. Radiology. 2007;244(2):356–78.
6. Weismann C, Mayr C, Egger H, Auer A. Breast sonography: 2D, 3D, 4D ultrasound or elastography? Breast Care (Basel). 2011;6(2):98–103.
7. Gonzaga MA. How accurate is ultrasound in evaluating palpable breast masses? Pan Afr Med J. 2010;7:1.
8. Abdullah N, Mesurolle B, El-Khoury M, Kao E. Breast imaging reporting and data system lexicon for US: interobserver agreement for assessment of breast masses. Radiology. 2009;252(3):665–72.
9. Vreugdenburg TD, Willis CD, Mundy L, Hiller JE. A systematic review of elastography, electrical impedance scanning, and digital infrared thermography for breast cancer screening and diagnosis. Breast Cancer Res Treat. 2013;137(3):665–76.
10. Sadigh G, Carlos RC, Neal CH, Dwamena BA. Accuracy of quantitative ultrasound elastography for differentiation of malignant and benign breast abnormalities: a meta-analysis. Breast Cancer Res Treat. 2012;134(3):923–31.
11. Barr RG. Sonographic breast elastography: a primer. J Ultrasound Med. 2012;31(5):773–83.
12. Evans A, Whelehan P, Thomson K, Brauer K, Jordan L, Purdie C, et al. Differentiating benign from malignant solid breast masses: value of shear wave elastography according to lesion stiffness combined with greyscale ultrasound according to BI-RADS classification. Br J Cancer. 2012;107(2):224–9.
13. Berg WA, Cosgrove DO, Dore CJ, Schafer FK, Svensson WE, Hooley RJ, et al. Shear-wave elastography improves the specificity of breast US: the BE1 multinational study of 939 masses. Radiology. 2012;262(2):435–49.
14. Chang JM, Moon WK, Cho N, Kim SJ. Breast mass evaluation: factors influencing the quality of US elastography. Radiology. 2011;259(1):59–64.
15. Xydeas T, Siegmann K, Sinkus R, Krainick-Strobel U, Miller S, Claussen CD. Magnetic resonance elastography of the breast: correlation of signal intensity data with viscoelastic properties. Invest Radiol. 2005;40(7):412–20.
16. Krouskop TA, Wheeler TM, Kallel F, Garra BS, Hall T. Elastic moduli of breast and prostate tissues under compression. Ultrason Imaging. 1998;20(4):260–74.
17. Muthupillai R, Lomas DJ, Rossman PJ, Greenleaf JF, Manduca A, Ehman RL. Magnetic resonance elastography by direct visualization of propagating acoustic strain waves. Science. 1995;269(5232):1854–7.
18. Lawrence AJ, Muthupillai R, Rossman PJ, Smith JA, Manduca A, Ehman RL, editors. Magnetic resonance elastography of the breast: preliminary experience. In Proceedings of the international society for magnetic resonance in medicine. Sydney, Australia: International Society for Magnetic Resonance in Medicine; 1998.
19. McKnight AL, Kugel JL, Rossman PJ, Manduca A, Hartmann LC, Ehman RL. MR elastography of breast

cancer: preliminary results. Am J Roentgenol (AJR). 2002;178(6):1411–7.

20. Sinkus R, Lorenzen J, Schrader D, Lorenzen M, Dargatz M, Holz D. High-resolution tensor MR elastography for breast tumour detection. Phys Med Biol. 2000;45(6):1649–64.

21. Sinkus R, Tanter M, Catheline S, Lorenzen J, Kuhl C, Sondermann E, et al. Imaging anisotropic and viscous properties of breast tissue by magnetic resonance-elastography. Magn Reson Med. 2005;53(2):372–87.

22. Sinkus R, Siegmann K, Xydeas T, Tanter M, Claussen C, Fink M. MR elastography of breast lesions: understanding the solid/liquid duality can improve the specificity of contrast-enhanced MR mammography. Magn Reson Med. 2007;58(6):1135–44.

23. Siegmann KC, Xydeas T, Sinkus R, Kraemer B, Vogel U, Claussen CD. Diagnostic value of MR elastography in addition to contrast-enhanced MR imaging of the breast-initial clinical results. Eur Radiol. 2010;20(2):318–25.

24. Barr RG, Zhang Z. Effects of precompression on elasticity imaging of the breast: development of a clinically useful semiquantitative method of precompression assessment. J Ultrasound Med. 2012; 31(6):895–902.

25. Chen J, Glaser KJ, Stinson EG, Kugel JL, Ehman RL, editors. Non-contact driver system for MR elastography of the breast. In: Proceedings of the international society for magnetic resonance in medicine. Montreal, Canada: International Society for Magnetic Resonance in Medicine; 2011.

26. Chen J, Grimm RC, Glaser KJ, Kugel JL, Pelletier KM, Ehman RL, editors. Improved noncontact 3-dimensional breast MR elastography. In: Proceedings of the international society for magnetic resonance in medicine. Melbourne, Australia: International Society for Magnetic Resonance in Medicine; 2012.

27. Glaser KJ, editor. MR elastography inversions without phase unwrapping. In: Proceedings of the international society for magnetic resonance in medicine. Hawaii, USA: International Society for Magnetic Resonance in Medicine; 2009.

28. Manduca A, Oliphant TE, Dresner MA, Mahowald JL, Kruse SA, Amromin E, et al. Magnetic resonance elastography: non-invasive mapping of tissue elasticity. Med Image Anal. 2001;5(4):237–54.

29. Chen J, Brandt KR, Ghosh K, Grimm RC, Glaser KJ, Kugel JL, et al., editors. Noncompressive MR elastography of breasts. Proceedings of the international society for magnetic resonance in medicine. Salt Lake City, USA: International Society for Magnetic Resonance in Medicine; 2013.

30. Shah SH, Hayes PC, Allan PL, Nicoll J, Finlayson ND. Measurement of spleen size and its relation to hypersplenism and portal hemodynamics in portal hypertension due to hepatic cirrhosis. Am J Gastroenterol. 1996;91(12):2580–3.

31. Bolognesi M, Merkel C, Sacerdoti D, Nava V, Gatta A. Role of spleen enlargement in cirrhosis with portal hypertension. Dig Liver Dis. 2002;34(2):144–50.

32. Talwalkar JA, Yin M, Venkatesh SK, Rossman PJ, Grimm RC, Manduca A, et al. Feasibility of in vivo MR elastographic splenic stiffness measurements in the assessment of portal hypertension. Am J Roentgenol. 2009;193:122–7.

33. Mannelli L, Godfrey E, Joubert I, Patterson AJ, Graves MJ, Gallagher FA, et al. MR elastography: spleen stiffness measurements in healthy volunteers-preliminary experience. Am J Roentgenol. 2010; 195:387–92.

34. Chen J, Stanley D, Glaser K, Yin M, Rossman P, Ehman R, editors. Ergonomic flexible drivers for hepatic MR elastography. Stockholm, Sweden: Annual meeting of international society for magnetic resonance in medicine; 2010.

35. Nedredal GI, Yin M, McKenzie T, Lillegard J, Luebke-Wheeler J, Talwalkar JA, et al. Portal hypertension correlates with splenic stiffness as measured with MR elastography. J Magn Reson Imaging. 2011;34:79–87.

36. Yin M, Kolipaka A, Woodrum DA, Glaser KJ, Romano AJ, Manduca A, et al. Hepatic and splenic stiffness augmentation assessed with MR elastography in an in vivo porcine portal hypertension model. J Magn Reson Imaging. 2013;38(4):809–15.

37. Morisaka H, Motosugi U, Ichikawa S, Sano K, Ichikawa T, Enomoto N. Association of splenic MR elastographic findings with gastroesophageal varices in patients with chronic liver disease. J Magn Reson Imaging. 2013. doi: 10.1002/jmri.24505. [Epub ahead of print].

38. Shin SU, Lee JM, Yu MH, Yoon JH, Han JK, Choi BI, et al. Prediction of esophageal varices in patients with cirrhosis: usefulness of three-dimensional MR elastography with echo-planar imaging technique. Radiology. 2014; 272(1):143–53.

39. Ronot M, Lambert S, Elkrief L, Doblas S, Rautou PE, Castera L, et al. Assessment of portal hypertension and high-risk oesophageal varices with liver and spleen three-dimensional multifrequency MR elastography in liver cirrhosis. Eur Radiol. 2014;24:1394–402.

40. Shi Y, Glaser KJ, Venkatesh SK, Ben-Abraham EI, Ehman RL. Feasibility of using 3D MR elastography to determine pancreatic stiffness in healthy volunteers. J Magn Reson Imaging. 2014 Feb 5. doi: 10.1002/jmri.24572. [Epub ahead of print]

41. Yin M, Venkatesh SK, Grimm RC, Rossman PJ, Manduca A, Ehman RL, editors. Assessment of the pancreas with MR elastography. Proceedings of the international society for magnetic resonance in medicine. Toronto, Canada; 2008.

42. Venkatesh SK, Yin M, Grimm RC, Rossman P, Ehman RL, editors. MR elastography of the kidneys: preliminary results. 16th annual meeting ISMRM, Toronto. 2008.

43. Rouviere O, Souchon R, Pagnoux G, Menager J-M, Chapelon JY. Magnetic resonance elastography of the kidneys: feasibility and reproducibility in young healthy adults. J Magn Reson Imaging. 2011;34: 880–6.

44. Lerman LO, Bentley MD, Bell MR, Rumberger JA, Romero JC. Quantitation of the in vivo kidney volume with cine computed tomography. Invest Radiol. 1990;25(11):1206–11.

45. Warner L, Yin M, Glaser KJ, Woollard JA, Carrascal CA, Korsmo MJ, et al. Noninvasive in vivo assessment of renal tissue elasticity during graded renal ischemia using MR elastography. Invest Radiol. 2011;46(8):509–14.

46. Shah N, Kruse S, Lager D, Farell-Baril G, Lieske J, King B, et al. Evaluation of renal parenchymal disease in a rat model with magnetic resonance elastography. Magn Reson Med. 2004;52(1):56–64.

47. Bensamoun SF, Robert L, Leclerc GE, Debernard L, Charleux F. Stiffness imaging of the kidney and adjacent abdominal tissues measured simultaneously using magnetic resonance elastography. Clin Imaging. 2011;35:284–7.

48. Korsmo MJ, Ebrahimi B, Eirin A, Woollard JR, Krier JD, Crane JA, et al. Magnetic resonance elastography noninvasively detects in vivo renal medullary fibrosis secondary to swine renal artery stenosis. Invest Radiol. 2013;48(2):61–8.

49. Lee CU, Glockner JF, Glaser KJ, Yin M, Chen J, Kawashima A, et al. MR elastography in renal transplant patients and correlation with renal allograft biopsy: a feasibility study. Acad Radiol. 2012;19(7):834–41.

50. Streitberger KJ, Guo J, Tzschätzsch H, Hirsch S, Fischer T, Braun J, et al. High-resolution mechanical imaging of the kidney. J Biomech. 2014;47(3):639–44.

51. Venkatesh SK, Wang G, Thamboo TP, Liu E, Siew EPY, Anantram V. MR elastography of renal transplants: correlating stiffness with interstitial fibrosis and tubular atrophy. In: Proceedings of 20th annual meeting ISMRM. Melbourne2012, p. 4062. 2012.

52. Robboy SJ, Bentley RC, Butnor K, Anderson MC. Pathology and pathophysiology of uterine smooth-muscle tumors. Environ Health Perspect. 2000;108 Suppl 5:779–84.

53. Solomon LA, Schimp VL, Ali-Fehmi R, Diamond MP, Munkarah AR. Clinical update of smooth muscle tumors of the uterus. J Minim Invasive Gynecol. 2005;12(5):401–8.

54. Walker CL, Stewart EA. Uterine fibroids: the elephant in the room. Science. 2005;308(5728):1589–92.

55. Stewart EA, Taran FA, Chen J, Gostout BS, Woodrum DA, Felmlee JP, et al. Magnetic resonance elastography of uterine leiomyomas: a feasibility study. Fertil Steril. 2011;95(1):281–4.

56. Hesley G, Gorny K, Woodrum D. MR Elastography following Focused Ultrasound Treatment of Uterine Fibroids. Paper presented at: Focused ultrasound Therapy-2nd European Symposium; 2013 October 10-11; Rome, Italy

57. Kolipaka A, Woodrum DA, Araoz PA, Ehman RL. MR elastography of the in vivo abdominal aorta: a feasibility study for comparing aortic stiffness between hypertensives and normotensives. J Magn Reson Imaging. 2012;35(3):582–6.

58. Woodrum DA, Romano AJ, Lerman A, Pandya UH, Brosh D, Rossman PJ, et al. Vascular wall elasticity measurement by magnetic resonance imaging. Magn Reson Med. 2006;56:593–600.

59. Bahn MM, Brennan MD, Bahn RS, Dean DS, Kugel JL, Ehman RL. Development and application of magnetic resonance elastography of the normal and pathological thyroid gland in vivo. J Magn Reson Imaging. 2009;30:1151–4.

60. Cheng S, Gandevia SC, Green M, Sinkus R, Bilston LE. Viscoelastic properties of the tongue and soft palate using MR elastography. J Biomech. 2011;44(3): 450–4.

61. Kemper J, Sinkus R, Lorenzen J, Nolte-Ernsting C, Stork A, Adam G. MR elastography of the prostate: initial in vivo application. RoFo; Fortschritte auf dem Gebiete der Rontgenstrahlen und der Nuklearmedizin. 2004;176(8):1094–9.

62. Arani A, Da Rosa M, Ramsay E, Plewes DB, Haider MA, Chopra R. Incorporating endorectal MR elastography into multi-parametric MRI for prostate cancer imaging: initial feasibility in volunteers. J Magn Reson Imaging. 2013;38:1251–60.

63. Arani A, Plewes D, Chopra R. Transurethral prostate magnetic resonance elastography: prospective imaging requirements. Magn Reson Med. 2011;65: 340–9.

Index

S.K. Venkatesh and R.L. Ehman (eds.), *Magnetic Resonance Elastography*,
DOI 10.1007/978-1-4939-1575-0, © Springer Science+Business Media New York 2014